WHY HUMANS LIKE TO CRY

Tragedy, Evolution,
and the Brain

MICHAEL TRIMBLE

OXFORD
UNIVERSITY PRESS

SCENE 1

There Is a Miracle

My eyes glaze over the letters I've written to my parents and Heather. I walk to the kitchen and grab a serrated knife. A box cutter would do the trick, but I have to work with what I've got. Jerry Springer is on television. *One show I will not miss.* I look closely at the knife. I sit down on my satellite-dish chair. I feel each scallop of the knife edge wondering what it will be like when I make the first deep cut.

I flip over my arm so that my wrist is facing the ceiling. I stare at the knife. *Just do it, Greta.* I'm scared. *God, please help me make the right decision. What do you want for me?* For several minutes I sit in deliberation. *Wake up, stupid! You don't have to do this. You can have a chance at life. You don't have to die.* A voice of reason and hope, and, most likely, fear washes over me. I rocket myself out of the satellite dish and put the knife away. I'm not going to die. I'm about to start living! For a brief moment, my depression lifts.

I celebrate with a pizza, a cheesecake, and a six-pack of Diet Coke.

With two more days to go before rehab, I'm determined to consume my last hurrahs. I call my friend Scarlett and make plans to meet an hour later at an expensive French restaurant. We drink bottle after bottle of pinot noir. At the end of the meal, I am stuffed.

"I'm so full," I whisper to Scarlett.

"Just go purge in the bathroom real quick."

"Yeah, but I feel bad. The bill is like $500."

"Who cares? It's no big deal." I had my answer.

"Excuse me. I need to use the restroom." With that, I was off to relieve myself.

A few minutes later, I came back to the table feeling two pounds lighter.

"Okay, let's go to a bar," Scarlett says.

"I'm in." I say. It's my last night. I have to live it up.

BEEP, BEEP, BEEP, beep, beeeeeep!!!

My eyes pop open from a deep alcohol-induced sleep and I realize the horn of the taxi that's barking is for me. I look out my window and see a black Town Car waiting. Shit. I cannot miss my plane. I have a nasty hangover. The clock reads 7:30 AM. *Fuck.* Panic-stricken, I jump out of bed, stumble across the room to my closet, and throw on a sweatshirt and an already too-tight pair of jeans. I'm not sure, but I think I'm still drunk. My flight leaves at 8:45 AM.

I grab my large duffel that I thankfully packed yesterday. That might have been the only act of self-care I have done in the last three weeks. I barrel down five flights of stairs and throw myself into the Town Car.

"Sorry, my alarm clock didn't go off," I say apologetically. "La Guardia, please." The cab driver's eyes glare at me in annoyance through the rearview mirror. *Gimme a break. I'm still drunk and I'm leaving for rehab.*

The cabby's New York driving style of pushing back-and-forth on the gas pedal makes my stomach flip. I really can't afford to do anymore puking at this point. As we race over the Triborough Bridge, leaving behind rows of sandwiched skyscrapers, I can't help but wonder what my life will look like in the next twenty-eight days. For the price tag, it better get a hell of a lot better.

I arrive at the airport just in time to check in and board the plane. My hangover has taken on a Xanax-like quality—masking my usual terror of flying. I'm what flight attendants call a nervous flyer, a.k.a. control freak. If I'm moderately afraid, then I might tell the person next to me that I hate to fly, forcing this anonymous stranger to comfort me because I lack the ability to self-soothe. If there is turbulence, forget it. I will put my head down on the tray table, pray, or even cry. Today, I don't care if the plane crashes. At least it would be over.

By the time we land, my hangover has cleared. I walk to the baggage claim and wait for my duffel bag. My nausea has been replaced by nervousness. This is it. I'm going to rehab. An older man walks toward me. He's the type I'd imagine wearing Wrangler jeans and drinking a Pabst Blue Ribbon beer next to his cactus on the front porch of a double-wide trailer.

"Are you Greta Gleissner?" My disheveled appearance clearly spells: "Girl in Need of Rehab."

"Yes. I'm just waiting for my bag."

"Hi, I'm Joe," he says, shaking my hand.

"Hi," I say, trying not to look embarrassed. I suppose I don't appear nearly as damaged as those who have alcohol and drug problems. They probably come staggering in drunk or crazy paranoid from cocaine-induced psychosis.

I sit in the front seat of a white van with Joe. It's the end of June, so it's pretty hot. We make our way out of the airport and onto destination Sierra Tucson. I would've preferred to come to rehab really skinny like my last hospitalization. At least then I could feel like I'd earned my bed.

Sometime after leaving the airport, Joe turns into a convenience store and parks the car.

"Well, this is the last stop. Buy whatever you want to eat or drink now. It's your last chance." I wonder if he does this for the alcoholics he picks up, allowing them to chug a six-pack in the backseat in case a cop drives by.

I think about everything I won't be able to eat during the next twenty-eight days, maybe forever. I'm not hungry, for once. I already feel so fat. I think I've done enough damage. I grab a Diet Coke.

Arizona is bare. No grass. Dirt and cacti everywhere. I don't understand how people live here. As we make our way up the long stretch to the rehab, I gaze out the window and wonder what my new life will be like. We pass a road sign that says, Expect a Miracle. Hope radiates through my body. A miracle is exactly what I need.

WE PULL UP to the circle drive to the nirvana of rehabs. Did Joe accidentally bring me to the luxury spa next door? Who would've thought that in the middle of the Santa Catalina Mountains, among wide-open spaces filled with cacti and shrubbery, would be such a beautiful place? The campus is large with Adobe-style buildings, serene walkways with sitting areas, a swimming pool, and a horse stable. There are people walking around; one guy sits on a large rock, in a meditative stare. Rodin's *The Thinker* statue comes to mind. Maybe I, too, will sit on a rock and contemplate my internal struggles. The exterior facade certainly takes away from what this place is: a psychiatric hospital.

I take a seat moments before a tall, robust man with a jolly face walks up to me.

"Hi, my name is Tim McCleod. I'm the alumni coordinator at Sierra Tucson."

"Hi, I'm Greta."

"Where did you come in from?"

"New York."

"Oh, what part?"

"The city. Manhattan," I say, feeling the need to clarify.

"So are you from New York City?"

"No, I'm from Kansas City."

"Get out of here. I'm from KC!"

I can't believe it. The first nice person I meet is from my hometown. I'm in the right place.

"Greta, come with me, please," a woman calls out to me.

"Bye, Greta," Tim says. "I'll see you later."

I'm handed a gray binder that outlines where and what I will be doing for the next twenty-eight days. First, I'm about to meet with the nurse. This will be one of several assessments that make up Sierra Tucson's Medical Assessment and Stabilization Unit (MAS). I will reside in MAS until I have completed my history and physical, psychiatric evaluation, Beck depression inventory, psychosocial interview, and a therapeutic recreation assessment.

I take a seat. The nurse sits across from me. She's sifting through my chart.

"I'm Barbara," she says, holding out her hand.

"Hi," I say, shaking her hand.

"So, Greta. You're here for bulimia."

"Yes," I say. After thirteen years, I'm not embarrassed by my eating disorder. Bulimia is a part of me.

"You've been struggling for a long time I see," she says. Barbara looks at me, her eyebrows and lips frown empathically. The sincerity in her eyes forces my tears to rise.

"Yes, I have." Beneath my tears, all I can think about is how fat I must look to Barbara.

"Sierra Tucson is a wonderful place. You're going to get well here. We treat the whole person: the mind, body, spirit, and emotions." Whole person? I only relate to the labels of dancer and bulimic. Everything else I'm not so sure about.

"Let me have your right arm. I'm going to test your blood pressure."

Barbara wraps the adult-sized cuff around my arm. The last time I had a cuff around my arm I was at my skinniest state and I was hoping that my compulsive exercising and constant purging would have granted me the eating-disordered prize of child-sized cuff. I was able to fit into some child-sized clothes back then, which was an anorexic accomplishment in my sick mind.

I haven't weighed myself in weeks, but I know there will be no chance in hell of a child-sized cuff today, or in my future. I never will return to my skinniest weight. That sucks.

"118 over 72," she says.

It figures. I puke thirty times a day, and I have normal blood pressure. My potassium will probably be normal, too. I should be happy about this, but damn, anorexics have all the glory. They wear child-sized cuffs and children's clothing, receive one-on-one monitoring, attention, and admiration by the nurses, are forced to drink weight-gaining Ensures, are extensively covered by insurance, and more. Anorexics are noticed, whereas the average-weight bulimic receives nothing extra because we don't look sick.

Just like with dance, I'm great at my eating disorder, but I don't have the discipline to rise fully to the top of the eating-disorder ranks. I know how sick it is. I'm hoping my opinion will change in the next twenty-eight days. *Expect a Miracle.*

The nurse takes blood work and then sends me on my way. Next, an intake worker completes a psychosocial interview with me. The usual questions: How long have you had your eating disorder? How many times do you purge? Physical health problems? Do you have a psychiatric history? Any problems with substances? Self-injurious behaviors? Family history of addictions?

By the time I finish with my admission assessments, it is late in the afternoon. I'm not hungry because I ate a Cinnabon cinnamon roll at La Guardia airport—without purging—something I never do. In my binder, I find my MAS female meal plan. The meal plan exchange system brings to mind an image of Richard Simmons, in his bright red tank top and matching shorts, promoting his *Deal a Meal* card system on an infomercial. His diet is similar to Sierra Tucson's in that you deal yourself from a meal card so many servings of carbs, protein, fruit, veggies, and fat per day. As a teen, unable to order from infomercials, I'd created my own meal cards. I was sure they would be the answer for my weight-loss. *Wrong.*

After the assessment, I walk to the Sierra Tucson bookstore to buy a pair of sweats. I have to be comfortable if I'm going to keep all of my food down. I admire all of the recovery books, imagining that someday I might have my own memoir of recovery gracing their shelves. *Let's just get through one day, Greta.*

The first meal I eat here is dinner. Even though I haven't met with the dietician, I know from reading my gray binder the exchanges I'm allowed.

DINNER
1 fruit
2 oz. protein
1 starch
1 cup raw vegetables
½ cup cooked vegetables
1 fat

Contradictory thoughts and feelings flood my brain. I feel excited by the meal plan, thinking that I might lose weight. As I look through the list of exchanges, however, I also have feelings of projected depravation. I can't survive on one starch! *Yes you can, fat ass.*

There are also rules that we have to abide by for our meals.

- *Be on time. Breakfast: 6:30 AM; Lunch: 11:30 AM; Dinner: 5:00 PM.*
- *Get 100 percent of your meal plan on your tray before checking in.*
- *You must eat at meal monitoring table unless eating with family.*
- *You must have one hot entrée per day (either lunch or dinner). If you don't pick a hot meal, the dietician will pick one for you at the following lunch.*
- *Peanut butter is allowed only at breakfast or scheduled snack.*
- *Limit Equal use to two per meal.*
- *Crackers and rice cakes are not allowed starches at mealtime. You can eat them for your scheduled snack.*
- *Confront eating disorder behavior.*
- *No events (i.e., massage) at 5:00 PM.*
- *Eat only from your tray, no sharing.*
- *Eat slowly; pay attention to taste, texture, and temperature of food.*
- *Must eat 100 percent of your meal.*
- *If on anorexic protocol, you will drink Ensure if you eat 100 percent.*
- *If on bulimic protocol, you will stay with staff for thirty minutes after finishing your meal.*

The rules don't surprise me. They are pretty standard. I look down at my watch: 4:40 PM. I walk out to the Butt Hutt, the designated area for smoking—the only addiction they allow to be active on the grounds—and have a quick smoke before dinner. It has an area for several people to sit. Add a bartender, some liquor, and a tumbler glass filled with floral umbrella toothpicks, and this hut would be a tiki bar. I smoke two cigarettes in a row. Nicotine is my only source of solace.

The dining hall is large with soaring ceilings, a wall of windows, and enough tables and chairs to accompany the seventy patients who are staying at Sierra Tucson. On the right side of the room, several tables have been pushed together to make one long table. Eating disorder patients are first to circulate through the cafeteria-style line. I join them.

"Hi, I'm Greta," I say to the huddle of women. I try to maintain eye contact, but the bright lights in the dining hall spotlighting my face forces my eyes away from the group. *They're looking at your scars. You're so ugly.*

The garden-variety eating disorder stereotyped women look beneath my shameful appearance and provide a warm welcome. There's Hannah, the hollow-cheeked aged anorexic; Brittany, the overweight binge eater; Sue, the older, chronic dieter with hypothyroidism; Leslie, the thin marathon runner with exercise bulimia. A few look perfectly put together, like a trio of Chi Omega sorority sisters who secretly binge on and then purge pizza and beer in the dormitory bathroom stalls.

"We eat at what everyone calls the Sideshow table," Brittany says, pointing to a row of tables pushed together.

"They call it that because you never know what will go on at the table," Sue says.

The kitchen staff opens the doors.

"Just follow me," Brittany says. I mimic Brittany, grabbing a tray, drink glass, and silverware. I study my meal card. Staff members stand behind a glass counter waiting to take my order. I'm no stranger to this type of setup. I have binged at many cafeterias. The difference here is that each entrée and side dish has listed the number of starches, fat, vegetables, protein, fruit, and dairy. I choose chicken, mashed potatoes, and mixed vegetables. I grab a yogurt for my dairy. Tonight is a dessert night, so I am handed a piece of pineapple cake. I guess dessert doesn't count in the exchanges. I'm really not too keen on cake with fruit on it, but at this point it's the only way to have sugar.

I sit down at the table with my tray. I'm the newest eating disorder patient. Lenore, the dietician, is the tray checker. She looks at my meal card and my tray to make sure it aligns.

"Just eat what you can," she says to me. "I know this is your first meal. You'll get an individualized meal plan when I meet with you for our nutrition assessment."

"Okay, thanks." I look around at the other tables where newly sober addicts sit chatting and laughing, eating as much as they can since they no longer have their vices. A hiss of anxiety surrounds our table. Women stare at their trays, some with shaky hands as they pick up their forks, unsure as to how they'll survive the meal. Others contemplate in which order to eat their food or into how many pieces to cut their protein. Lenore watches closely for food rituals. I assess my tray. I like to save the best for last, so I plan to eat the vegetables first, followed by the chicken, potatoes, yogurt, and dessert.

"Hi, I'm Fran," the woman next to me says. Fran is probably midforties and very thin. Her shoulder-length brown frizzy hair goes in all directions. Beneath her messy exterior, I can see a classiness that probably exists when she is not in rehab. Outside of these walls, I imagine she lives somewhere sunny because of her tan. She probably has a housekeeper, maybe a personal chef or personal shopper. She's very thin, but with a lot of muscle, so I'm guessing she's bulimic. She has a certain carefree quality about her that lacks the rigidity of the anorexics I have seen.

Fran starts jabbering almost incoherently in some sort of double-talk, hidden-meaning type of language. The reality hits me that while this looks like a spa that Oprah might go to, it is a psychiatric hospital. Fran keeps eyeing my pineapple cake.

"Hey, are you going to eat that?" Fran asks, pointing to my cake.

"No." I don't have to eat everything today. It's my first meal. I'm just trying to get through my chicken and potatoes. I don't want to think about the dessert.

"I'll take it when Lenore isn't looking. Don't say anything." While Sierra Tucson doesn't look like a hospital, the sneaky behaviors are familiar.

"Okay." I'll go with it. I don't want to make enemies on my first day.

"Jane," Lenore says, "Try to take bigger bites." Lenore isn't paying too much attention to us at the moment. She is monitoring Jane, an anorexic at the other end of the table, who's been eating her salad for the last fifteen minutes. Taking miniscule bites is a food ritual, which is prohibited.

A few minutes pass and my cake has been abducted by Fran's bulimic bear claw. Some of the girls notice, but don't say anything. A couple girls roll their eyes. I guess they're used to Fran. She babbles on throughout the meal, incoherently at times. By the time I finish my mixed veggies and chicken, I decide she's insane.

Lenore scrutinizes our trays to make sure we each have eaten our food before jotting notes down in a pad. I wonder if she will make us show her our napkins or check to see if food is in our pockets. "Marsha?" Lenore calls out.

"Yes?" Marsha is a bubbly, freckle-faced chubby girl. She looks like a camp counselor.

"You will be Greta's bathroom buddy." This is nothing new to me. At the last inpatient unit I was at, bathrooms were locked at all times, and you had to have a nurse watch you piss and shit.

"Hi, Greta," Marsha says as she waves to me from the other end of the table. "We have to stay together for thirty minutes." I can tell by Marsha's happy glow that she has been here for a while.

"So, how is it here?" I ask.

"It's a great place, really. The staff is awesome. The food is good. You can talk to a therapist 24/7." Her words make me feel at ease.

"How long have you been here?"

"About two weeks. Let me know if you want to go to the bathroom. Do you want to go out to the Butt Hutt?

"Sure." I always smoke cigarettes after I eat. Besides purging, it's the only thing that calms me.

The Butt Hutt is filled with newly recovering drug addicts, from crack to meth to alcohol. More introductions take place. I imagine the groups of addicts judging me. *If you'd just gotten skinny before you came, it would be so much better, Greta. I hate myself.* Sierra Tucson is known for attracting wealthy types or celebrities. I'm just fortunate that my parents have good jobs and know how to manage their money.

"Hi, I'm Teresa," a young blond says, sitting amongst a tight-knit clan. I feel insecure. I can tell this place has its cliques. I'm the new kid in school and she's in the popular crowd.

"I'm Greta," I say. "What are you here for?" *I'm so fat and ugly next to her. She's skinny. I used to be that skinny. Now I have to romp around in cut off sweats and sweatshirts in the Arizona heat because I feel so disgustingly fat.*

"Cocaine." No wonder she's so thin.

"Hi, I'm Melissa." Melissa sat with me at the Sideshow table. "We may be in the same primary group." Each patient has one core group they attend for therapy each day.

"Nice to meet you," I say, hoping she can't see the insecurity all over my face. I introduce myself to the others. I can feel the quad sizing me up, staring at my acne scars and fat.

Following dinner, I listen to a recovery speaker share his experience, strength, and hope. He discusses how his addiction led him to steal, how it ruined his relationships, and caused him to be homeless. I can identify. I've stolen from my parents, from grocery stores, and from friends to maintain my food addiction. I barely have any friends; and since I was evicted, I have no home. There is no difference between him and me. We are both low-bottom junkies.

"Who's that?" I ask my bathroom buddy.

"That's Vicki, one of the evening-weekend therapists. A lot of people have problems with her. She's too bossy."

Vicki is in her midforties and her energy is captivating. She has spunk, but her directive manner could come off as a nag for those who don't like authority figures. Vicki walks toward me.

"Hi, I'm Vicki. You're new today?" I shake Vicki's hand. Vicki's gentle green eyes reveal sincerity beneath her tough exterior.

"I'm Greta," I say. As she smiles back at me, adrenaline warms my veins. I can already feel the emotional suction toward her. Within thirty seconds, I decide I need her. She will be the one who will take care of me while I'm here. *Here you go again, Greta.*

"I'm here on the evenings and weekends. If you need anything, at any time of the night, let me or one of the other therapists know. You can talk to a therapist anytime night or day."

"Okay, thanks." *Don't you worry, Vicki. You can count on me.*

After the meeting, I go off to my room in the MAS area. I throw myself on my bed and reach for my *Just for Today* food addiction daily reader and prayer on the nightstand. It has been months since I've read it or the dozens of Overeaters Anonymous phone numbers I have scribbled inside the front and back covers.

Just then it hits me. I did not purge today. I can't believe it—*I did not purge*! If I wasn't so exhausted I'd jump up and down on my bed the way I used to as a kid at Grandma Sally's house. I have one day of abstinence. I have had one day before, but this time it feels different. I feel like I'm not white-knuckling it. I feel like I just might surrender.

MY DAY BEGINS at 6:00 AM. Before my eyes even open my hand is fondling my stomach to check for signs of possible weight loss from my one day without purging. *You're still fat, Greta.* I pop out of bed and take myself to the dining room for my first breakfast. I'm hungry. I do the same routine as I did at dinner, choosing my allotted portions of whole grain cereal, yogurt, a peach, and one tablespoon of peanut butter. I grab two Equals and a big cup of decaf coffee (caffeine isn't allowed).

I go to the Sideshow table and set my tray down.

"Hi, I'm Jill," says a young, vibrant woman wearing a Sierra Tucson staff badge. "You must be Greta. I alternate checking meals with Lenore."

"Oh, okay," I say. "Here's my tray." Jill eye's my tray, comparing it to the Sierra Tucson beginner's meal plan, before giving me the green light to sit down. Jill's athletic build and career choice makes me wonder if she's in recovery herself.

Breakfast goes well. It's a relatively easy meal for me. Before racing off to my first therapy group, I go to the Butt Hutt to smoke away any feelings I have about eating my breakfast. People sit around reminiscing, telling war stories, just the way they'd do in a bar, only without the cocktails. Instead of the usual pickup lines and small talk, the casual chat at this joint is: Where are you from? and What's your drug of choice?

I land in my primary eating disorders group right on time. Eight clients sit in a circle looking like plated lobsters about to be cracked open for dinner. We stare at the therapist, anticipating the utensils and shell crackers she'll use to pick out every morsel inside of us—the sweet emotional meat of why and what we hide behind our shell—until there's nothing left and we're totally raw.

I know all of the women by face since I've sat with them for two meals, but Linda, the therapist, wants us to introduce ourselves. Therapy begins and I pull out my journal to take notes. I want to take in as much knowledge as I can.

Linda, the therapist, looks in my direction. *She must be thinking what a good patient I am for journaling.*

"We don't write during group," she says.

"Oh, okay. Sorry." Shit. I fucked up already. Can't she see I am trying to be the perfect patient? *You stupid, ass-kissing piece of shit. You're so ridiculous!* I'm embarrassed, but I cover it up. This, of course, is the very thing I need to process. My perfectionism and inability to hear anything that is potentially negative.

"Today, I want to talk a bit about how your eating disorder has worked in your life," Linda begins. "What is it that makes it so hard to give up?"

Group members start sharing their reasons.

- It's the only thing I can do well.
- I don't deserve anything else.
- I want to punish myself.
- It's the one way to get the attention I need.

I raise my hand.

"I have this empty pit inside that I'm always trying to fill. It's like I'm always hungry."

"Uh-huh. What are you hungry for?" Linda asks me.

"Well, I guess part of it is that I want acceptance. When I was skinnier I always thought that I needed to be skinny because I had to make up for how ugly I was."

"That sounds so painful, Greta. Is there anything about yourself that you like?"

"Not really. Only my eyes."

The group gives me supportive feedback.

"Greta, I wish you could see what I see when I look at you. You're so kindhearted."

"You're so hard on yourself."

"I feel the same way you do, but I don't see why you feel like I do when you're so pretty."

At the end of the group, Linda makes an announcement.

"Tomorrow, I want you to write a letter to yourself about your eating disorder. So start brainstorming about what you want to say, and we will spend tomorrow's session writing and discussing it." Anxiety flutters around my stomach. I have a lot to say about my eating disorder.

My primary eating disorders group is the first of dozens of groups packed into my twenty-eight-day stay at Sierra Tucson. Educational lectures, family therapy, healthy sexuality/LGBT, exercise, nutrition, grief group, codependency, and a host of experiential therapy sessions, such as equine therapy, climbing on a rock wall, EMDR (eye movement desensitization and reprocessing), and acupuncture. My eyes glaze over at the schedule, exhausted by the amount of time I have to focus on myself.

I make it through lunch. If I felt this physically full in NYC, I'd be looking for a bathroom. My arm wraps itself around my stomach trying to protect me from feeling the discomfort. Still, I feel like I am missing something. This is the very feeling that I always try to numb out and eat over. Bingeing and purging are the only source of comfort and self-soothing I know. Here, I can't do either of those. I just have to sit with my feelings. Ugh.

In the evening at orientation group, we focus on the first step in Overeaters Anonymous: "We are powerless over food and our lives have become unmanageable." John, one of the evening/weekend therapists, is the leader for the group. He's a tall buff Hispanic with an intense stare. He could be a bouncer in a nightclub. He's probably a recovering addict. Crack and alcohol, I'm guessing.

"Tonight we are going to talk about powerlessness and unmanageability," John says to the room of twenty. A few whisper to one another and giggle quietly. My eye contact and attentiveness mask the boredom I feel. "Hey, eyes open, Rich," John says, singling out a young surfer type. "What's powerlessness?" John asks the group.

Rich wears adidas flip-flops, long loose shorts, and a tank top. He opens his eyes and looks at John with attitude, as if he has disrupted his nap. "I don't really think I'm powerless over marijuana."

"I see," John says. "Who would like to tell me what powerlessness is?"

Hands begin to raise. People around the room describe their version of powerlessness: "Can't stop yourself." "Have to use even though you don't want to." "Tried not to use and did anyway." "Tried controlled use." "Told yourself you'd only have one and didn't."

I'm certainly powerless. My daily experience with food spells out powerless.

"So, Rich?" John begins. "Can you identify with anything people are saying?"

"Yeah . . . I guess." Rich is more awake, but his frown says he doesn't like what he hears.

"In what way?" John isn't letting up. He is focusing on Rich, but knows that others in the room probably feel the same way.

"Well, I spent two years sitting two inches from the television, smoking out of my bong. I felt like I had to use every second of the day."

I can identify with Rich. If I'm not using, then I'm thinking about using. *You're a total junkie, Greta.*

I'm pooped. After the meeting, I fall onto my bed so exhausted that I don't even read my nightly meditation book. I'm astonished that I've made it through one full day of treatment and two nights without bingeing and purging. *There is a miracle.*

MANY APPOINTMENTS TAKE place during the first week.

"Get on the scale. Backwards, please." Lenore says. My first weigh-in. Weigh-ins are Tuesday and Fridays at 6:15 AM. The treatment team suggests blind weights so we don't freak out for the twenty-eight days. Seeing the number is one of my biggest triggers. She writes the number down on her pad. My eyes wander to the notepad. My 20/20 vision sees my weight: 146. Highest ever. Panic hits me in the gut. *Holy shit, I didn't know I was that big.* Since being weighed at Radio City, I have gained ten pounds.

A gentle wave of apathy washes over me, soothing my panic. *Who cares how much you weigh? You're in treatment for God's sake.* The number 146 flashing in my head interrupts my carefree attitude. Fear and disgust creep back in and soon I'm consumed by depression. My stomach suddenly feels full. I don't want to eat this fucking breakfast. My brain pushes the play button on the usual *fat-piece-of-shit-ugly-face-big-nose-worth-nothing* negative tape I verbally abuse myself with daily. The part of me that wants to recover talks back to the hateful voices in my head. *Trust the process, Greta.*

I go to my primary eating disorders group, ready to write a letter to myself about my eating disorder. Linda the therapist details the procedure.

"I want you to use your name in every sentence," she begins. Think about how the eating disorder has impacted your life. At the end, give a solution to what you will do to help yourself. You have about twenty minutes and then we will discuss."

We start writing our letters, hurriedly, trying to capture everything. My perfectionism makes it hard for me to write on the spot. One by one, we read our letters aloud.

Dear Greta,

Here you go again. You ate the breakfast you planned and then you had to have one more cup of coffee. And one more cup of oatmeal. Greta, you fool yourself everyday. Greta, you tell yourself that you can handle it, that you won't purge, you'll just eat nothing the rest of the day. Or, Greta, you will convince yourself that you really do need the extra food. Well, Greta, that is a bunch of shit. You, Greta, cannot handle any amount of extra food, let alone a food plan. Greta, you always purge. Greta, you need to get real. Greta, you eat or drink liquids just so you can purge. And Greta, you're not anorexic anymore—you're not going to wither away by eating a healthy portion of food. And now, Greta, you've just finished purging and are ready to make a day out of bingeing and purging.

Greta, no! Greta, this is going to stop now. You have a choice. Greta, know that if you give into your disease you will spend at least $50 of your own money, order food on your mom's credit card, you will be physically worn out,

emotionally drained, and will feel depressed. You will feel
guilty for giving into the behavior and getting nothing done
with your day. And guess what, Greta? The feelings you are
trying to avoid—they will still be there. Greta, you know
from many years of experience that the feelings always come
back. Greta, your way doesn't work anymore. You can't do
the same thing over and over again expecting a different
result. So, Greta, this minute, you are going to change. You
are going to a meeting today. When you want to pick up the
food or purge, you, Greta, will pick up the phone. You will
start on an exercise plan again. Greta, not to lose weight,
like in the past, but for physical as well as mental health.
Greta, you have the willingness to change the things you can
and this is one of them.

I feel a little lighter once I read my letter aloud. I can see how I am separate from the eating disorder. I have an illness; but I am not my eating disorder.

My life in rehab consists of eating and processing my feelings in therapy. If only life could be this simple. For hours each day, I think about what keeps me in my eating disorder and what needs to change.

By Wednesday of my first week, I'm cleared for medical stabilization and move into a building called Morning Star. The name reminds me of my favorite veggie burgers. My room is classy, with accommodations that could easily compete with a room at the Marriot Midtown East. I put my clothes away and set my daily meditation books on the nightstand as part of my spiritual growth regimen. A note sits on the nightstand: *Welcome to Sierra Tucson, Greta. I look forward to being your roommate. Peace and love, Sandy.* She sounds nice.

Besides groups, more appointments fill my week. I have an exercise assessment with Jill. We go over my exercise history and what

I have currently been doing for exercise. That's easy—nothing. I haven't danced for at least a couple of months because of my depression and lack of money. Jill and I instantly connect, as she used to be a competitive swimmer and had an eating disorder. She even had an article written about her in *Sports Illustrated*. She uses her past struggles to help others, which is exactly what I want to do.

I meet with the psychiatrist, Dr. B.

"Come in, come in. You must be Greta. I'm Dr. B."

"Hi," I say, holding out my hand for the umpteenth therapeutic handshake. Dr. B. has tight frizzy curls and she's not the most attractive, but from the lecture she gave about the brain and addiction, I can tell she's brilliant. In addition to being a psychiatrist and an eating disorder specialist, I've heard that she's a biochemist. You don't need looks when you have her mind.

Dr. B. flips through my chart and conducts a psychiatric evaluation, asking me about past medications, diagnoses, occupation, social history, etc.

"What do you see as some of your issues you'd like to work on while you're here?" Her beady brown eyes study me.

"Well, obviously, I want to work on my eating. I want to figure out how to be happy with myself in dance, and not worry so much about what others think."

"It says here that you have sexuality issues?"

"Yeah," I say. My eyes fall to the floor. *I hope she's not homophobic. She has to be open as a professional, right? Is Arizona a liberal state? She probably thinks I'm a disgusting person.* "I think I'm a lesbian, but I'm not sure fully if it's because of past issues I have had with men or because I have mother issues. It's confusing to know what the truth is." Tears of shame trickle down my cheeks. *I cannot be gay.*

The last group of the day is a community meeting, where we do something called *care*frontations, which means a gentle way to confront each other. I have not done anything like this before. One by one, people in the community start sharing whatever beef they have with another patient. Whoever is being confronted can only answer: It fits. It doesn't fit. And thank you. The point is to hear the observations of what people see in you without becoming defensive. I don't want anyone to call me out on what they see. Note to self: *Don't piss off anyone.* The rest of my evening is consumed with dinner and relapse prevention.

I'm not looking forward to the LGBT sexuality group at all. Fear and shame embodies me when I think about or say the word gay. Lesbian has a less shameful ring to it. The LGBT group is small. In the room are the common stereotypes—the flamboyant gay man, the androgynous butch woman—but no one looks like me. I'm not butch enough to look gay, and I'm not girly enough or interested in men enough to be straight. I seem to not fit into any group, gay or straight. Maybe I will just keep trying to like men. It seems easier to live in the world like that. Easier for my parents, but maybe not for me.

The weekend is less stringent, but still packed with activities and groups. The most exciting part about Sunday is brunch. The chef makes personal egg requests. We even are allowed cinnamon rolls. Looking at my tray filled with an omelet and a cinnamon roll, I feel excited. My mouth waters in anticipation. I check in with Lenore and sit down. I start eating, enjoying every bite. Laura, a young emaciated woman, is not happy. She does not like brunch, today.

"Laura, can you get through the meal? Do you think you can try to take some bites?"

"No, I can't do it." Laura is becoming more anxious by the second.

"Laura, I will need you to eat, or else you will have to be supplemented." Supplements like Ensure are scary. Who wants to drink creamy, fattening liquid? I decide to focus on myself and on my food.

Suddenly, Laura stands up and slams her tray filled with food against the wall. "I hate this food! I don't want to eat it!!" Vanilla yogurt trickles down the wall. Chunks of cinnamon roll and omelet cling to the wall like an installation piece, which could've almost been appropriate had this been art therapy. I hate to admit it, but if no one was looking I'd be tempted to eat the remnants of cinnamon roll. Smashed or not, it still looks yummy.

Her tray tantrum has silenced the room. Even some of the drug and alcohol addicts who've given away their luxury cars or who've prostituted themselves when they've felt this level of desperation and panic seem stunned. Put a pile of cocaine on the table and tell them not to snort it, and I'm sure trays would be flying all over the room. I can empathize, but my first instinct is to laugh. *C'mon, Laura. That's a creative antic to getting out of brunch.* I mean, where am I? Where do you see this? Oh, yeah. At a psychiatric hospital.

Laura runs out of the room with a counselor right behind her. I do feel badly for her. I'm sure that if I were in her condition I'd throw my tray, too. *You're not that bad, Greta.* Within minutes, the wall is cleaned up and the room has resumed equilibrium.

The Sideshow table has lived up to its name today.

DURING THE SECOND week, I hear whispers swirling around campus: *Did you hear who's coming here? Can you believe it?* Soon enough, a musician from a popular boy band arrives. Of course, he doesn't stay in Morning Star. He stays in the other residence, Crescent Moon. When I see him, I immediately feel embarrassed. I want to treat him like a regular person, but it's hard not to be affected by his fame. I'm dying to talk to him because his bandmate's wife went to my Kansas high school and she's a former Rockette, too.

A girl nicknamed Princess is going nuts. "Oh my God. He's here! He's here!" The staff reminds us of the confidentiality policy and how we cannot call family and friends because the paparazzi might find him. For now, I'm going to act like he's not here. I have to focus on myself.

IN MY PRIMARY group, we are working on a "Tough Love" list.

"This week we will be preparing for next week's Family Week," Linda says. "To prepare, I want you to write down a series of statements to your family members who are coming, in the form of 'When you *(fill in the blank)*, I feel *(fill in the blank)*.' Write as many statements as you can. Then you will pick three that you will read to your parents next week during carefrontations."

I've never been good at confronting my parents about anything. My hand automatically attaches itself to my stomach. *I feel fat.* Linda hands us a blank worksheet to start brainstorming. Twenty minutes later, I have my list. Besides my mom and dad, Kent is coming, but I don't have any carefrontations for him. He's physically and emotionally always been there for me.

I'm nervous for family week. I feel guilty that I'm even at Sierra Tucson. I can only imagine my parent's carefrontations. Maybe some good will come out of it.

Expect a Miracle.

Instead of going to the fitness center for exercise group, we meet in a much larger room, similar in size to a gymnasium.

"Today, we will be climbing the rock wall," Rick, the facilitator says.

I'm excited. I'd always heard about these rock walls, but had never actually seen one.

"The point of the exercise is to build trust. You will have two people holding the cords, supporting you as you climb up the wall."

He gives us instructions on how we climb up the wall and how to support others. A recovering alcoholic volunteers to go first.

"You said you have a hard time asking for help, so I thought this would help," the instructor says, handing the man a blindfold to wear.

I'm up next. The instructor hands me a twenty-pound backpack to symbolize the emotional weight I carry around daily.

I crawl up the rock wall one leg at a time, like the tarantula I saw crawling up the side of the Butt Hutt the other night. My back hurts from the added weight.

"How does it feel to carry that weight around, Greta?" he asks, as I'm scaling the wall.

"Not good," I say, about halfway up to the top. "It's really heavy."

"As you're climbing, you can decide when you want to take that weight off of your back. When you're ready, drop them."

When I was back down to ground level, the instructor had a few comments for me.

"Greta, I noticed you dropped the weights after I gave you permission. Do you know why you didn't drop them sooner or why you didn't hold on any longer?"

"Well, I was tired. I wanted to drop them earlier, but I felt like you might be disappointed if I did it earlier."

"I see. Do you always do as you're told in an effort to not disappoint?"

"Yeah, I guess. I feel like there's a lot of pressure in keeping people happy."

"What else did the weight on your back represent?"

"Guilt and shame." I am guilty of so much regarding my eating disorder. All the lies I've told and the amount of money and time I've wasted. The shame comes just for being me. Besides being able to dance there isn't anything I see good about me. I hate what I look like. I'm probably gay. I have so much regret for decisions I've made. All of it adds up to one big ball of shame. *Disgusting.*

"How important do you think it will be to your recovery to stop trying to please others and get rid of guilt and shame?"

"Probably really important." I know it is. I do feel lighter just from doing the exercise and talking about it.

In between therapy sessions, I've been bonding with my eating disorders group members, mostly hanging with Fran and Melissa. All of us come from different backgrounds, different situations, and different family dynamics. Yet we all have the same core feelings. We all experience the battle between the eating disorder voice and the healthy voice, unsure as to who will win on any given day. Even though externally we might appear like normally adjusted people, for some reason, all of us feel undeserving and we emotionally binge on self-hatred. I look at my peers and wonder how they could possibly hate themselves. I can see their strengths. I'd like to be able to see my own.

Fran and I overhear the Princess talking to her friends from home on the phone about Boy Band Man. She'd already snooped through his mail, so this is another huge confidentiality breach. We decide to tell one of the therapists. The next morning, Princess is on a plane home for the indiscretion. Thank God. She's what my Grandma Sally calls a Nosey Nedderman. We don't need her busy-body antics or her higher-than-thou attitude.

IN MY NEXT eating disorders group, we review our carefrontation lists. After writing and rewriting, my list is complete. Linda asks that we pick three statements that we will read aloud to our family members. I'd rather not say any of them.

Mom's List
- *When you ask me if I'm going to put on makeup, and are so quick to pay for medicines and procedures to fix my face, and then turn around and say how pretty my friends and*

cousins are, I feel ugly, ashamed, inadequate, embarrassed,
and believe you think I am ugly.
- *When you don't have anything to talk about except the*
  weather, my schedule, and why I'm not dancing anymore,
  I feel angry, sad, lonely, and hurt.
- *When you are uncomfortable hugging me, saying I love*
  you, nurturing me in any physical way, other than giving
  me money, I feel rejected, hurt, inadequate, and guilty.

Dad's List
- *When you've been unsupportive of my dancing throughout*
  my life, until a few years ago, and make jokes about serious
  situations, like my car accident and eating disorder, I feel
  angry, guilty, sad, and hurt.
- *When I come to your house and I ask you the same ques-*
  tion over and over because you are too busy watching TV
  and aren't listening to me, I feel unimportant, inadequate,
  and sad.
- *When you feel upset because I have to divide my holiday*
  time and weekends between you and my mother, I feel
  angry and guilty.

There is a part of me that feels so guilty for having the thoughts
and feelings listed on my carefrontation. I know that anything my
parents have said or haven't said isn't out of malicious intent. They've
proven they love me with the amount of money, clothes, and constant bailing me out of financial situations. They'd do anything to
help me. At the same time, I don't understand why it is so difficult to
show me affection or say they are proud of me.

I breeze through lunch. I must be on what the rooms of Overeaters Anonymous call a "pink cloud," which tells me I'm floating

through the process too smoothly. The meals are easy; my obsession is nonexistent. I almost feel fantastic at times. Life is grand. The hard part about the pink cloud is that at some point you fall off of it and bounce back to a depressive painful reality. A similar feeling could be if I thought all of my pants were a size two, only to realize that for months I'd misread the labels and they were really size twelve.

Family Week

'm half way through the program and I can barely eat brunch in the dining room because I am anticipating my parents' arrival. They were never involved in this way in any of my previous treatments. I'm not sure what to expect or how they're going to feel. My mom, dad, and Kent are coming. Nervousness overcomes me thinking about my mom and dad being in the same room together. The last time was when I was in rehab three years earlier. My father couldn't even bring his eyes off the floor to look at my mother.

They never communicate. Since their divorce, I still serve as the communication liaison for scheduling how much time I spend with each of them. Of course, neither party is ever satisfied with the schedule I come up with because they both feel they don't have enough time with me. Only my eating disorder brings my parents together.

Now they'll be in the same room together for therapy all day long. Geesh. Lucky for me their therapy is in a location separate from mine for much of the week.

Sierra Tucson's family program is just as well planned as all of the addiction programs. They have family groups during the day while I remain in my groups. We come together only to do family therapy in our primary eating disorder group and for eating meals. We have instructions about how and what we can communicate with our families during their visit. We can only talk to them about serious issues when we are in the therapy room. At meals, we have to keep the conversation light. The family members stick to the same rules that we do at meals regarding the type of communication.

When my parents arrive, I greet them as if they're in town to watch me perform at Radio City. I see my mom and Kent first. Kent's eyes glisten with empathic tears as he comes close to give me a hug. My mom gives me a stiff double pat on the back. Though my mom smiles at me, tension exists behind her eyes. I'm sure she'd rather be anywhere but here. I mean, who really wants to rehash family business with your exhusband and bulimic daughter at Family Week?

I see my dad across the room, off to the side. I give him a hug and he joins the three of us, though he stands on the outskirts. I feel guilty. My dad always feels cheated by the lack of time he has with me, and it's happening already. I silently repeat a mantra I've learned in the last week. "Your parents can emotionally take care of themselves." We stand around looking at each other uncomfortably, unsure as to what to say.

The first part of the week is the same for me, as they are participating in their own family therapy sessions. I hear through some other family members that the therapist and group hammered my mom and dad for the way they handled the divorce. I don't know the specifics, but apparently their issues were spotlighted on the therapeutic stage for all to witness. One performance I'm sure they could've lived without.

I eat my meals with my parents in the dining room, alternating the time I spend with them, just as we do on the outside. We make stupid conversation since we are all trying to abide by the rules. "No food or weight talk." "No family issues talk." "Save it for group." After lunch on the third day, my dad makes a bold statement.

"Greta, after the first day of therapy, I was going to leave. I was ready to get on a plane and go home."

"Really, why?" I say.

"It was really not a good time. I called Lavada and she helped convince me to stay. I thought I'd better stay for you." My stepmother is the voice of reason.

"Well, I'm glad you did." Immediately I'm wracked with more guilt. If only I didn't have this eating disorder, then none of us would have to be here and my parents wouldn't have to feel vulnerable.

My mom and Kent have a different reaction.

"Greta, I don't know how you do this therapy all day long," my mom says. "I can't believe we're finished for the day and you still have four more hours left. It's exhausting."

The next day is the big day. We start the first of two days of carefrontations, which are done during my primary eating disorders group. My mom, dad, and Kent have the chance to speak first.

The setup is the normal therapeutic circle, but it is much larger to accommodate the family members. Inside the circle two chairs sit: one for the family member and one for the patient. Just seeing the empty chairs in the center of the circle fills me with fear. My parents have shown great concern for my behavior over the years, but the primary emotion is anger. I don't want to hear it. I know I need to, but I don't want to. Several family members go before we are called.

"Okay, Greta," Linda says. "Why don't we do your family's carefrontations?"

Kent volunteers to be first. I stand up from my chair with reluctance and head to the center of the circle and take a seat. Kent sits opposite of me. He has tears in his eyes already.

"Greta, when you use my credit card numbers to buy food, I feel angry, worried, and scared."

His voice shakes and has a higher pitch as he cries through his carefrontation. Pressure builds in my head. I begin to cry. I'm not allowed to respond. All I'm allowed to do is listen and process. He starts the second carefrontation.

"Greta, when I find cups full of vomit under your bed, I feel angry, sad, and worried." He begins the third. "Greta, when I see you not take care of yourself, I feel sad and angry." Kent looks up at me and whispers, "I love you." I say it back as he stands up and returns to the outer circle.

"Gayle," Linda says to my mother. "You can come up next.

My mother replaces Kent and takes a seat. Instinctively, her back sits up straight and her legs cross, giving her a nice posture and leg line. She would fit in well posing with the Rockette line for photographs. Her eyes pierce mine before looking down at her paper of carefrontations. I can't gauge her temperature. She's either about to cry or is really pissed.

"Greta, when you spend all of your money on food and use our credit cards to buy food, I feel angry and upset." I could've guessed that was going to be one of hers. I've been a financial wreck and ridiculously irresponsible because of my eating disorder.

"Greta, when you wear your Rockette jacket on the street in New York City and don't do your hair or wear makeup, I feel angry and embarrassed."

My mother starts rattling off the third carefrontation, but I've shut down emotionally. All I can hear is the second carefrontation. My heart is racing. I want to stand up and throw my chair across the room. Doesn't she understand that this, this precise dynamic of worrying about how I appear, and potentially how it will reflect on her, is a huge problem in our relationship? This is the exact dynamic that has existed since at least eighth grade when she was fixated on my wearing foundation to cover up my acne.

I want to crawl out of my skin. I want to tell her how none of this appearance bullshit matters. Besides anger, I feel embarrassed. Doesn't she realize that everyone is analyzing what she's saying and that to mention something so appearance-driven in an eating disorder group will only implicate her as a key player of how I ended up at Sierra Tucson? Doesn't she understand that the purpose of this whole experience is to be able to get in touch on an emotional level?

Maybe she should explore why she is unable to physically touch me and say normal mom things, like "I love you," or "You're special," or "You're fine just the way you are." For someone as smart and multi-talented as my mother, it amazes me how clueless she is in this area. Her behavior is equally as infuriating and embarrassing as me showing my bare ugly face while wearing my Rockette jacket strolling in Rockefeller Center.

I want to say all of these things, but all I'm allowed to do is listen until tomorrow, when I say my carefrontations. I'm not looking at my mother at this point. My eyes search for an exit, shifting from the floor to my therapist.

"Thank you, Gayle. Rob, you can come up next."

My dad sits down on the chair. God, I can't take much more of this. Outside of these walls, I'd be on my way to every donut pub and fast-food restaurant, attempting to emotionally bury these feelings. But here all I can do is take it. An acute awareness of the way my stomach is touching my shorts invades my space. I'm feeling so fat. *Fat's not a feeling, Greta. Shut up, it is too!*

My dad begins. "Greta, when I had to find out from Dr. Oxler that you had an eating disorder because you and your mother hid it from me, I feel hurt, angry, and left out." Twelve years later, he is still holding onto that. I get it, but at the same time, why doesn't he look at his part in the situation. If he'd been able to act like an adult and communicate with my mother in matters that concern me, maybe he would've known sooner. As a sixteen year-old, I certainly wasn't capable of telling him my shameful secret.

Furthermore, after he did know, his discomfort with the subject made him say remarks at the dinner table like, "You're not going to throw that up are you?" I know this was his way of trying to provide comic relief to a serious situation. My dad's intent was not malicious, just misguided. To some extent, I can say the same for my mother. They both have always wanted the best for me.

Through my intense emotions, I'm still trying to make sense of all of their comments and see the silver lining. My mother stirred up the most emotions. Maybe her seemingly superficial focus on my appearance is really her saying she is worried that I hate myself so much that I don't bother caring about how I look. Maybe I'm giving her too much credit. I'm not sure.

The next day I'm ready to present my carefrontations.

"Okay, Greta. You and your mom can come to the circle." Instead of cues to stand front and center onstage, I'm now commanded to take my place front and center in the circle. This gig is not a performance. There are no costumes, makeup, or false eyelashes. Just raw emotions.

My stomach tap-dances with nerves. Prior to this experience I wouldn't have had the courage to come face-to-face to my parents and actually say how I feel. Today, though, I'm stronger. I feel the support of the room. *It doesn't matter what the outcome is. Just take the next right action.*

I read my carefrontations, unable to look my parents in the eyes. I've been waiting so long to say I'm hurt, I'm angry, I feel rejected and unloved by you. Yet as I read the rehearsed lines, I feel guilt and shame, reminded of how much they've given me, and recognizing how hard it is to be a parent. Surprisingly, a hint of stoicism overrides my nerves and I say most of my carefrontations without emotion.

The next therapy session, the four of us meet privately with my therapist to discuss aftercare plans. I believe my aftercare is going back to New York to see Heather and attend Overeaters Anonymous meetings. I'm surprised when I hear Linda say that I need to continue treatment at a step-down facility. What? I need to get back to New York. I have to get in shape for the Rockettes. My parents and I look at Linda like a deer in headlights.

"Why does she need to go somewhere else after this?" my dad asks.

"What our experience has shown is that a twenty-eight day program is not enough, especially for Greta. Her eating disorder history is long and severe, and so to have the most favorable outcome, we recommend a step-down facility where she can transition back into life."

"How long will she need to go somewhere?" my mom asks. I imagine she is thinking about how this will interfere with the Rockettes. I am thinking this, too.

"Well, probably sixty to ninety days," Linda says. Sixty days. I start doing calculations. That would be the end of September. Rehearsals start at the beginning of October this year. I'd be cutting it close, but it is possible.

"Greta, you look preoccupied. Do you want to continue treatment?"

"Yes, I'd want to. I am just thinking about Radio City, the timing, and the money. This place is so expensive. I feel bad that my parents will have to pay more." That's really what this is about. I don't care so much about going back to dance. I'm guessing that my parents' biggest concern is that they have to pay for extra treatment. I mean, I should be better by now.

"Yeah, what happens with New York and dance? She's supposed to be dancing at Radio City in the fall," my mom says. I catch my dad staring at my mom, annoyed that she's concerned with this.

"Well, the one thing we know is that anything you put before your recovery you lose," Linda starts. "I know Greta has already lost so much. I also know that dance has been, in part, one of her triggers for her eating disorder, even though she loves it." *She's got a point.* "Greta, I hear that you're worried about the money. If your parents are able to pay for it, would you go?" *What choice do they have now? Of course they're going to do anything for their daughter.*

"Yes," I say, hesitantly.

"So you're saying that for Greta to have the most success, she needs to do this extra treatment?" Kent asks.

"Yes," Linda says.

"Greta, what's the problem?" my dad asks.

"I just feel bad about the money. I've already spent and wasted so much."

"Greta, you have to do what's necessary to get better." His assertion lets me know that he is more worried about my health than what it will cost him, which soothes my feelings of guilt for the interim.

My mom, dad, and Kent stare at me.

"Okay, I'll go."

I have to choose between a facility in Montenido, California, and two in Tampa, Florida. The one in California is ridiculously expensive, similar to living at Sierra Tucson for three months. Even though my parents could afford it, I won't let them. Unsure if I wanted to live for the next sixty to ninety days in a small Victorian house with just eight or nine women, I decide on Turning Point of Tampa, which is coed, coaddicted. While there may only be eight or ten people with eating disorders, the treatment center has probably twenty to thirty people, which is much more of a community than the other option.

On the last day of family week, the families and patients all gather in one room. The facilitator asks us each to say three things we like about each other. As we go around, my dad surprises me. He says he likes my face, which is the very thing I despise the most. He looks right into my eyes and says it. Wow. Tears begin to flow. I've never heard him or my mom mention that they like my face. He really thinks I'm okay just as I am.

"I want to make an announcement to the group," my dad says to the room. My dad's voice is loud. I can't possibly imagine what he is going to say. "I just want to thank everyone here. You know, I came into this week and immediately wanted to go home. I almost left, but then decided to stay for my daughter, Greta. I just want to say that I'm so glad I stayed. I've learned a lot and met some great people. This experience has changed my life! So, thank you!" I feel so proud as I smile up at my father.

My parents and I hug goodbye. Everything feels much lighter than five days earlier. We took care of some emotional business. Though my mom and I still have a lot to work out.

ON MY FINAL day of therapy, during grief group, I read a letter I wrote detailing the purposes dance has served in my life and what I miss about dancing. Surprisingly, the therapist asks if I'd like to dance in front of the group.

"Uh, I don't know . . . dance in front of the group?" I've performed on stages since I was three years old, but improvisation has never been my strong suit. I haven't danced in months and I'm so much fatter now. If I dance, people might think I'm not really that good. *You, a professional dancer? Sure you are.*

"Yes, Greta. It might be good to be connected to dance in a therapeutic setting."

"We support you, Greta," says one of the patients.

"Okay. I guess I could do a little something." I have my cutoff Sierra Tucson sweats and a T-shirt on, which is what I used to wear to class in New York.

I stand up, do a few pliés and some stretches, and prepare. The therapist has us move to a private hallway for extra space. Everyone is staring at me, waiting for me to perform. *Just dance like no one's watching you, Greta.* I start dancing my favorite lyrical style, doing kicks and pirouettes and jetés down the hallway, with only the music in my heart choreographing my moves.

At first, all I can think about is how the patients are probably judging me. But soon, I become lost in the movement. As I dance, tears trickle down my face in joyous remembrance of how I used to feel. The movement ignites my soul. This is a feeling that can come close to filling me up as much as food does. I miss this feeling. My tears turn from joy to sadness, and whip across my face as I turn, grieving the loss of who I was and who I may never be again.

A couple of minutes pass and I have to stop. I'm pooped. I have lost my stamina. I look around at my audience. Each has the same starry-eyed look. They're impressed by my abilities.

"You're amazing, Greta!" one says.

"Yeah, so beautiful," says another.

"How did it feel to dance, Greta?" the therapist asks.

"I felt a mixture of emotions. It's hard to imagine not dancing. I've always been a dancer. I guess I have to figure out who I am besides being a dancer and having an eating disorder. Which feels kind of scary, but exciting at the same time."

Twenty-eight days ago, I could never have danced in front of people without wondering what I looked like and how they were judging me. I finally feel free.

The evening before I leave, we do what's called a clap out. In an area called the Kiva, we stand in a circle and each throw a rock into a big dirt pit to symbolize something negative we want to leave at Sierra Tucson. One by one, the people who will be leaving prior to the next clap out yell out loud what they will leave behind. I know what I will say.

"I want to leave behind self-hatred, shame, and the feeling of inadequacy!" I yell. It feels good to say it out loud. I meant it. I want to leave it behind. I've been carrying it with me for way too long.

After we throw our stone, each of us stand by a huge rock in the middle of the Kiva, place both hands on it, and state something we'd like to take away from Sierra Tucson.

"I'd like to take self-love and self-acceptance from Sierra Tucson." As I say it, I feel the energy from the rock, imagining a hefty dose of self-love and self-acceptance radiating through my body. After the clap out, I feel more love and self-acceptance.

The next day, I pack all of my belongings. I climb back in the white van with Joe and we head for the airport. As we leave the property, I see the backside of the same sign I read the first time I drove into Sierra Tucson. In this direction, however, the sign reads differently: You Are a Miracle.

In this moment, I do feel like a miracle.

A Turning Point

I fly from Tucson to Atlanta. The Atlanta airport has taken on a new appearance. Before, I saw the airport through the lens of bulimia, looking at every kiosk, newsstand, and restaurant as my next binge. Now, the airport is just an airport. And I am just another traveler. Yes, I still smell the decadent scent of Cinnabon wafting through the terminal when I pass by. I'm not immune. But I'm not the same girl who, just twenty-eight days ago, had to binge before a flight or feel guilty for eating a package of delicious biscotti cookies. I'm in recovery now, and I think I'm beginning to resemble something close to normal. Hot damn! As quickly as I think of myself as normal, the thought of celebrating with a piece of cake reminds me, once again, that I am not normal. I am a food addict.

Food addict or not, I board the plane a changed woman. Instinctively, I cross my hands and lower my head to begin my plane-crash-prevention prayer ritual. *God, please let me be safe and don't let the plane crash.* I start to repeat the mantra, but then stop myself. *You don't have to do that anymore, Greta. You have faith now, remember?* Oh, yeah. *If I'm in fear, then I'm not in faith.* Of the many recovery clichés I've picked up at Sierra Tucson, this is, perhaps, the one I need to remember the most. And with that, I release my hands from the tight clasp. Seconds pass, and fear creeps in again as I survey all of the children on board the plane. *I'll be okay. God wouldn't take all of these children.*

I arrive safely in Tampa surrounded by large beautiful palm trees. It's hot, but nothing like Arizona.

Down at the baggage claim, a short beefy thirty-something man walks up to me.

"Are you Greta Gleissner?" His voice is coated with a two-pack-a-day nicotine habit.

"Yes, that's me," I say. I must still look like I belong in rehab for someone to pick me out so quickly.

"Hi, I'm Jerry. I'm a house manager at Turning Point."

I wait for my bag and then we hop in his black Ford Explorer.

"So, Sierra Tucson? How was it?" Jerry's flirty eyes and wet smile says he might be labeled as a "thirteenth-stepper," hitting on vulnerable newly recovering girls who'd like to replace a man for their substance of choice.

"Sierra Tucson was amazing. Life-changing."

"Yes, the country club of rehabs," Jerry says with a jovial growl just under his breath.

"Yes, it is. What's Turning Point like?"

"It's not Sierra Tucson."

Driving away from the airport, the sun beats down on my face. *Ah, hours of daily therapy in the sun. This is the life.* I remind myself that I'm not here because I chose to relocate to Florida for spa treatments and rejuvenation. I wanted to kill myself thirty short days ago. *Oh, that.*

"Does it have a bookstore?" I ask, thinking about if I'll be able to buy a pair of Turning Point sweats to match my Sierra Tucson sweats.

"You'll see," he says, with a husky laugh. We turn into a cul-de-sac. "Home sweet home."

My cheerful exterior masks my apprehension of living in one of the cottages on the circular patch of asphalt. There will be no bookstore or Turning Point sweats. At the very most, I might be provided with a Turning Point pen for my journal. That way when I write about how fat I feel and how I want to puke, the Turning Point logo will provide a Pavlovian reminder: *Don't relapse, Greta.* Turning Point embodies the "step down" concept. Though anywhere would be a step down from Sierra Tucson.

Nicknamed the Village, the complex houses two populations: eating disordered and chemically dependent. Patients are segregated by our addictions, and if someone is dually diagnosed then they are in the eating disorder houses. In the center of the cul-de-sac sits the building known as 4A. This is where house managers do their work, and where we receive our psychiatric medications and a weekly allowance, which we call "escrow money." It also has a bedroom for clients who break rules and can't be trusted.

Jerry and I enter 2A, my new home, which is connected with another eating disorder house, 2B. The house has a Floridian feel to it, with floral print couches and bluish carpeting. On the beige Formica kitchen countertops sit measuring cups and food scales. A bulletin board hangs with clients' weekly meal plans. No more chefs. Back to reality.

"Okay, I have to check your luggage."

He sifts through my duffel, pouring the contents of my makeup bag onto the bed in search of contraband. Just like at Sierra Tucson, he removes potential self-harm instruments, like fingernail clippers and eyebrow tweezers. Any liquids with alcohol are shoved into a quart-sized plastic Ziploc bag, which will be returned to me at discharge. Luckily, I didn't keep any of the mini-Snickers bars

Fran offered me for this next leg of my treatment journey. He rifles through my purse, wallet, and change purse.

The only other time I've been ransacked like this was at the Burbank Airport when flying to Kansas City. "Excuse, me, Miss," a mid-fifties Caucasian male says, flashing an LAPD badge in my direction.

"Uh . . . Yes?" I say, my insides quivering. I'm already guilty.

"I need you to show me your ticket and your identification, Miss." He tells me his name, but I only recall the words detective and special narcotics agent. *Shit.*

He is a plain-clothed undercover agent. Apparently, I look like a drug-trafficking derelict. Must be the grungy thrift store clothes my parents desperately wish I'd stop wearing. My fingers fumble through my wallet and pull out the requested documents.

The detective studies my driver's license and ticket.

"Okay, thank you," he says, handing my documentation back to me. "There's a woman we're looking for about your age who is carrying drugs with her, and you matched her description." *Great.*

After the requisite search, Jerry and I hop back into the Explorer, and drive to the Beaumont Center, a strip mall of ugly brown office buildings where I'll receive treatment. We enter the building that looks more like a place to complete your yearly taxes than it does to do the emotionally taxing work of treatment.

Employees all around glance up at me as I pass by. Some look up from their paperwork to notice me, the new client. Administrative staff members look up and give me a lukewarm smile. *Just another addict*, I imagine they're thinking.

The admissions coordinator takes down some basic information and then walks me to my next destination. Time feels rushed as I'm scooted from person to person to complete the admissions process. On the way to Bob Z., the intake therapist's office, Robin, the CEO, and nurse Ellen introduce themselves to me.

"Hi, Greta!" says Ellen, a middle-aged bubbly blond. "I'm getting with you later for an assessment."

"Oh, okay. Well, see you in a little while." She seems nice.

I continue walking through the building, peeking in the office doors of therapists and wondering who will be my therapist. We stop in front of Bob Z.'s office.

"Hi, Greta. I'm Bob. Nice to meet you." I shake his hand and say hello. Bob's salt-and pepper hair and matching mustache tells me he's in his late forties or early fifties. I wonder if he's in recovery. Why else would he be in this field? "We're going to do an intake, and then you'll meet with Ellen to discuss your health and the medications you are on."

Like at Sierra Tucson, he asks me questions about my eating disorder and substance use history. I begin to go through the entire story again. When my eating disorder started, the progression, blah, blah, blah.

"Greta, do you use any drugs or alcohol?" *Well, not in the last twenty-eight days.*

"I drink sometimes, but I don't use any drugs. The eating disorder is really my main thing."

"How often do you drink?

"On the weekends." *Except for that summer when I drank alone daily at the Markle Residence.*

"Okay. How much would you say you usually drink?" *Is this really necessary, Bob Z.? I told you I have an eating disorder, not an alcohol problem.*

"Well, maybe two to four glasses of wine. Sometimes less."

"How about any other substances?"

My mind traces back to the incident at the Burbank Airport. Before the special narcotics agent approached me, I had been at the trash can making sure my purse was free of drug paraphernalia before I entered security. Mr. LAPD didn't know about my minor eight-month, four-day-a-week crystal meth habit. My original plan was to smuggle meth into my luggage, but I was too chicken shit. How could I explain to my parents that I needed bail money for drug possession?

"Well, I smoked pot in high school and in my early twenties. And for about a year, I used crystal meth, to try to get away from the eating disorder." That's the truth. My eating disorder is so powerful that I thought if I did a drug known to make people waste away, a long-held anorexic fantasy of mine, then maybe I'd be distracted from my eating disorder and get skinny. It didn't work. I binged in between the days I got high. There was no weight loss. Strangely, the power of my eating disorder blocked the addictive progression of meth.

"Did you snort, smoke, or shoot it up?" He asks the questions with such ease, as if he's asking me whether I prefer Coke over Pepsi.

"At first, I snorted. In the end, I smoked it through a glass pipe." Bob Z. doesn't find any of this alarming. He's desensitized. *Another day, another drug.* "Even though it's such an addictive drug, I just put it down . . . and went right back to bingeing and purging." The truth was that once a drug scares me, I'm done with it. With meth I became too paranoid, and after thinking I was going to overdose one evening, I quit.

I say this trying to convince Bob Z., and perhaps myself, that I never really had any issues with substances. Yes, I abused highly addictive substances. But I wasn't *addicted*.

"When did you start using meth?" Bob Z. asks. He is honing in on this addictive phase of my life. *I'm here for bulimia, Bob.*

"Well, I was about twenty years old and living in Los Angeles when I first tried it. I started when I was working for this delivery company at a high-end Pan-Asian restaurant in Studio City. My manager, Joe, offered it to me.

"I see," Bob Z. says.

I snorted it casually for several months while Joe smoked it out of a glass pipe. One day I got curious.

"I want to try it," I said to Joe, enticed by the smell of the smoke coming out of his pipe in my bathroom.

"No, you'll get hooked." he replied. Note to self: *When a drug addict worries about your well-being, you may have a problem.*

"C'mon, just a puff," I said.

A month later, I was asking Joe for my own glass pipe to store in my bathroom drawer. Crystal meth hadn't yet gained the popularity with police-busting Midwest meth labs or the public. I didn't recognize the irony in my choice to store my pipe next to my toothbrush.

"I feel guilty, Greta," I remember Joe saying. "I'm getting you addicted."

"No you aren't," I replied. "I make my own choices." That was the most honest statement I had made in a long time.

Bob Z. seems intrigued by my meth adventures and wants to know more.

"So, you said you 'just stopped' using?"

"Yes. And went right back to the eating disorder."

We complete the intake, I move on to the next admissions station, Nurse Ellen. We quickly go over my health and psychotropic medications.

"You're very lucky," Ellen says, after I tell her that my only known health problems are some dental damage from vomiting. "It's amazing that you don't have any medical issues."

"I know," I say. The lack of medical problems is one of the reasons my eating disorder has persisted for so long. It adds to the denial. I don't see blood when I puke. I don't even have gastroesophageal reflux disease (GERD). *I am in perfect health.*

"Well, Greta. I'm going to take you to Lori now. She's the dietician. See you later," Ellen says.

"Oh, okay. Thanks, Ellen." Good, maybe I'll be able to eat lunch. It's close to 2:00 PM, and I still haven't eaten. I'm really hungry. *Maybe you'll get to skip lunch and drop a little weight.* My eating disorder mentality is still with me. Recovery reminders ring in my ears. *Hello, Greta . . . don't think that way. You're in recovery now!*

Next, I meet Lori, the dietician at Turning Point. She seems to be in her late thirties and has a normal body type, which I like. What person, especially someone with an eating disorder, would take dietetic advice from an overweight person? Lori has nice energy. On my last dietician meeting at Sierra Tucson, Lenore told me I had lost five pounds. Secretly, I'm hoping the adjustments Lori makes to my Sierra Tucson food plan will help me lose more weight. As much as I'd like to turn off the switch in my brain that prompts my hyperfocus on weight, I just can't seem to do it. I'm not going to be too hard on myself. My past dieticians have said that body image and focusing on weight is usually the last piece of the eating disorder to go.

"Well, I'm sure you're probably hungry. You've had a long trip. I wasn't sure what you'd like, so I had one of the other eating disorder clients put a lunch together for you."

"Oh, okay," I say, trying to not act as freaked out as I am by someone else preparing my food. *It's just one meal, Greta.*

"I've put your lunch in the employee break room, so follow me." I can't imagine what these strangers have made for my lunch. What if I don't like it? These people don't know what I like to eat.

I take a seat at the lunch table. The plate before me is as dull as the ambience of the break room. I'm supposed to eat two corn tortillas with oven-roasted turkey and mustard. I hate corn tortillas. Even though it has fat, I prefer mayonnaise, not mustard. At least they gave me some cheese for my fat intake. And I like raw carrots and fruit.

Sitting alone, in Tampa of all places, I'm feeling homesick for

Sierra Tucson. I miss the counselors, the Butt Hutt, the prepared meals. I have to learn how to live outside the cocoon of Sierra Tucson.

I finish my bland turkey wraps and head back to Lori's office.

"I know you just ate, so I'm not going to weigh you. I'll wait until we meet the next time."

"Great, because weighing on a full stomach is my worst nightmare." Lori nods empathetically. I hand her my meal card with my allotted servings.

"Wow, you eat three starches for breakfast and lunch?" Her astonishment lets me know she thinks three starches are too many. After all, I'm not a hypermetabolic anorexic who needs to eat three-thousand calories a day just to maintain. I'm a borderline chunky, untoned bulimic.

"Yes. After the first couple of weeks at Sierra, I was starving, so they upped my meal plan."

"How would you feel about eating two starches for lunch instead of three?"

"Yeah, that will be fine." *Maybe I will lose weight here.*

"So, tell me about your history." Lori seems shocked by my symptom use. "You have an extensive history."

"I know. I'm lucky to be alive. The only health issue I've had is a root canal."

"You really are lucky," Lori says. She's right. I've played Russian roulette with my body and my life for years. With all of the purging, I could've easily have had a gastric rupture, esophageal hemorrhage, or a heart attack. If I were a cat, I'm sure I'd have already zipped through my nine lives. Somehow God saved me. I must have a purpose.

There are major differences between Sierra Tucson and Turning Point's food plan. Turning Point adheres to an abstinent food plan, which means no sugar or white flour. Sierra Tucson believed in the normalization of eating. Turning Point considers condemned foods similar to how alcohol is for the alcoholic: One has an allergy that requires total abstinence. I'm not sure how I feel. In New York, Overeaters Anonymous defined abstinence as what you make it. You could eat chocolate daily and call yourself abstinent. I don't really think I have the allergy, as they say, but this type of abstinence certainly can't hurt me.

After our appointment, I follow other clients into a white van called the Druggy Buggy—the equivalent of the little bus that special-education kids ride in elementary school—that carts us back-and-forth to the Village. This is just one of the ways we've been infantilized. We also have house managers who babysit us during meals. And they watch us take our medicine, asking us to open wide after we finish making sure we swallowed the pills and are not holding them in our cheeks.

"Greta! Greta!" I hear from a distance. It's Teresa from Sierra Tucson. Teresa runs up and bear hugs me. We are suddenly best friends.

"Hi!" I say. I'm happy to know someone. "How is it here? It's so different from Tucson."

"It's good so far," she says, pulling out a cigarette. "Do you want one?"

"No, I have one," I say, lighting up a Marlboro Light.

"We do everything on our own. Chores, clean our rooms, and cook. The groups are good. It's more tough love than Sierra Tucson. I'm also in the eating disorders group."

"Oh, wow," I said. "You're okay with that? I know at Tucson, you didn't really want to discuss it."

"Yeah, well . . . I had to get honest with myself."

Soon others are coming up to the picnic tables to see me, the latest attraction. I receive a slew of introductions: Katie, Peggy, Steve, John, Jennifer, Andrea, Pauline. One tells me she's an alcoholic, another is a pothead, another smoked crack, and another is an alcoholic-bulimic. They no longer have last names. They've dropped their names for their disease labels they've married.

"Hi, I'm Pauline," one woman says, holding out her hand. Her short bowl haircut and androgenic face reminds me a little like the Pat character from Saturday Night Live. Only this Pauline is really skinny. I'm guessing she's around forty. Definitely a lesbian. "I'm in one of the eating disorder houses, 1A."

"And I'm Andrea," the other girl says. Andrea is normal weight to medical chart numbers, but probably a little chunky to my family's standards. Her long brunette hair and bright blue eyes give her femininity, but she has an inner-butch quality about her. My gaydar says she may be a lesbian or bisexual. *Maybe I'll be able to work out my sexuality issues here.*

"Oh, hey," I say to Pauline and Andrea. What are you here for?"

"Anorexia and alcohol," Pauline says. "What about you?"

"Bulimia mostly. I've had a few years where I teetered on the side of anorexia. But mostly, I just like to throw up."

"I'm here for bulimia and alcohol," Andrea says.

"How long have you guys been here?" I ask.

"I've been here a couple of months," Andrea says.

"Thirty-two days," Pauline says.

Two women walk toward our addicted clan.

"Hi, Greta. I'm Lisa and this is Maggie. We are your roommates in 2A."

"Nice to meet you," I say. I feel like that's all I've been saying all day."

"At 4:45 PM we start preparing dinner," my new roommate says. I look down at my watch. Fifteen minutes. That allows for at least two more smokes.

I can already tell that the picnic table is the Butt Hutt of Turning Point. We chain smoke and share war stories until dinner.

"So, are you from here, Pauline?"

"No, I'm from Texas. But I'm relocating to Tampa. They tell me I have to change everything, which means my environment."

"Yeah, the whole people, places, and things bit." I want to let Pauline know that I'm up on the recovery lingo. There's no way I'm trading in NYC for Tampa. Never. *I'm just here to get back on my feet. I'm different from you.*

My new house is the location where everyone with eating disorders consumes their meals. With a plethora of possible ritualistic food behaviors, this could be more entertaining than the Sideshow table at Tucson.

I follow the girls into the kitchen. I have to borrow people's food because Friday is food-shopping day. I feel weird asking anyone for help or borrowing their food, especially because I hate to share my food with anyone.

"Greta, what do you want to eat? You can have anything of mine," Teresa says.

"Thanks, so much. I know how it is to have to give your food to someone." Actually, I don't. I may steal food. But share? Never.

"Oh, I don't care." Teresa is like an atypical anorexic. She's more of a drug addict who has food issues. She doesn't appear to have the rigidity and control issues of the everyday anorexic. "I'll just eat what you're eating."

Teresa's measuring out one cup of mushrooms, onions, and peppers to put in her skillet.

"Here are the vegetables. I'm making extra rice, so you can use some."

"Great, thanks."

Jerry, the house manager, watches each one of us as we stuff our starches, vegetables, and fruit into measuring cups. I'm told there are two types of clients when it comes to measuring. The Fluffers and the Packers. The Fluffers, usually anorexics, put as little into their measuring cup; while people like myself, the Packers, try to shove a one and-a-half size portion of rice into an eight-ounce cup.

"Here you go," I say to Jerry, the house manager. "One cup of rice and one cup of vegetables." I place my grilled chicken breast on the food scale and weigh three ounces. "And here's my chicken."

"Okay," he says. "And how 'bout your fat and fruit?"

"I'm getting it now," I say. I measure out a serving of mayonnaise and grab one of Teresa's Red Delicious apples.

"Looks good," Jerry says.

Jerry, I, and the other seven eating disorder girls sit down at the table for dinner. I pick up my fork to dig in, but stop myself, realizing no one else is eating. Everyone has their hands out to the side waiting for their disordered neighbor to take hold. And then we begin.

"God, grant me the serenity, to accept the things I cannot change, the courage to change the things I can, and the wisdom to know the difference. Amen."

The eight of us start eating our allotted amounts, attempting to have regular conversations. Behind our small talk the white elephant in the room sits in a corner eating his can of abstinent circus peanuts. *You aren't a group of women meeting for a casual dinner. You all have mental disorders and do fucked up things with food.*

Obsessive and ritualistic behaviors are on the table. Some use a lot of condiments, like ten squirts of butter spray or an obscene amount of mustard on chicken. An anorexic puts black pepper all over her strawberries and blueberries. This is not a creative culinary move, like a frisée salad with watermelon, feta, and ground peppercorns. This is to punish herself and not enjoy it. Our behaviors will be dually noted in the house manager log. *5:42 PM: Katie pours pickle juice over her vegetables.*

Conversations are monitored also. We're not allowed to speak of dieting, calories, weight, or romanticize how great it felt to over-exercise, purge, or starve. Basically, nothing to trigger eating disordered thoughts or behaviors. As I learned at Sierra Tucson, if I don't entertain a thought, I won't act on it.

After dinner, we barrel into the Druggy Buggy and go to an Alcoholics Anonymous meeting. This is the first of my five required weekly meetings. Because Overeaters Anonymous meetings only meet a few days a week, we have to supplement the rest of our meetings with Alcoholics or Narcotics Anonymous meetings. I've never been to a public Alcoholics Anonymous meeting before. Coffee and cookies, two nonabstinent foods, are out. These are the newly recovering alcoholics' drugs of choice.

Suddenly, Sober Big Brother (a.k.a., my house manager) is watching me, so I choose decaf. I do manage to smuggle some packets of Equal into my pocket, which I will hide in my sock drawer later. Hoarding Equal is the only disordered thing I can do with food. I'm not sure why I need to get away with something. *Third Step, Greta. Make a decision to turn your will over to the care of God. Nope. Tomorrow maybe. But for now I need this.*

Whether in New York or Florida, the recovery clichés that buzz around the meeting room remain a constant. I wonder if I'm an alcoholic or drug addict as I listen to people tell war stories about what it was like before recovery. Addiction runs in my family. I've abused alcohol and drugs, but does that make me an addict? I began to use more frequently, but my tolerance never went up. At the height, I was still using moderately because of the fear-of-death factor.

I look around the room at addicts who are newly grateful and filled with their higher power. I suddenly feel empty and inadequate, which translates to *I feel fat.* I know these feelings well. No matter my substance of choice, these feelings are why I've earned this seat. I'm an addict.

After the meeting, I go to 4A to wait in line for my medication. "Come in," Hans says to me. "Greta, right?"

"Yes," I say. He reaches into a plastic bag labeled "Greta G." and doles out one tablet of Seroquel, which is a sleep aid. I've been on them since my first night at Sierra Tucson. I rarely have sleep issues, but because I'm no longer numbing my emotions with the eating disorder, I lay in bed tossing and turning, trying to escape the racing thoughts that surface as my mind flips back through time.

Sometimes my thoughts are about the mistakes I've made and the experiences I've missed out on. At times, I have sensory images from childhood, filled with constant screaming and feelings of not being safe. Other times, I flashback on the Northridge Earthquake or on a serious car wreck I caused when I was sixteen and my friend almost died. I've worked most of these issues out in therapy, but they still find a way to intrude my mind.

The Seroquel also helps block out what addicts call "drug dreams," only mine are filled with food. I dream of having just binged and purged, and wake up alarmed, wondering if in fact I had relapsed. You know your eating disorder is powerful when it gets into your subconscious.

I'm about to become a zombie. I swallow the pill with a mouthful of water. I open my mouth for Hans to look inside. I pass the pill-swallowing test and go back to my house.

"Okay, it's time for Meta," Jerry says, entering the house. Around 9:00 PM each evening, we eat a metabolic snack to regulate our blood sugar and metabolism. It consists of one starch, one dairy, and a fruit. Since Sierra Tucson, I've gotten used to eating a snack at night. I feel satisfied physically.

I haven't purged in twenty-nine days.

We have to go to bed by 11:00 PM. I drift off to sleep in a Seroquel haze. I'm not even disturbed by the flash of light that enters my room when house managers do their hourly checks. They want to make sure we are still in our beds and haven't made a run for a grocery store, bar, or drug hole.

Before I know it, a knock at the door wakes me. "Time to get up," the house manager says. It's half past six in the morning. I get out of bed, still woozy from my night medication, and slip my calloused feet into my adidas sandals to head to house 4A for my morning medication.

"Hi," I say. "I'm here for my meds."

"Are you Greta *Gleessner*?" The house manager asks. I don't know this one. This must be the overnight guy.

"It's *Glicener*," I say, phonetically correcting him, and then judging his arm full of tattoos, beer belly, and ratty beard. He probably drives a Harley to his Narcotics Anonymous meetings.

"I'm James," he says. He pulls out my medication from a bin marked with my name and hands me the 20 mg tablet of Lexapro and 50 mg of Topamax. Lexapro is an antidepressant and Topamax is a mood stabilizer. Recent research with bulimics showed positive outcomes for both drugs in regards to abating binge cravings.

I walk back to my house and prepare my breakfast. I am allotted two ounces of cereal instead of the two cups I usually equate to a serving. I have another half-cup of cottage cheese and a peach. We make decaf coffee in the morning and are allowed six artificial sweeteners a day. Since I usually use six in one big coffee, as well as an Equal in my cereal, I have to strategize. I set my bowl on the table just as the morning house manager, Ruthanne, sweeps into the room.

"All right, people, gentle breakfast!" Gentle breakfast refers to us eating in silence to help us concentrate on our food and the day we are about to begin. We are expected to think about our food as we eat it, enabling us to fully experience our meal. With her raspy, commanding voice, this frizzy-haired, mullet-wearing, fifty-something woman is anything but gentle.

We begin our gentle breakfast with a morning meditation. I'm beginning to wonder how I will survive breakfast in silence. Sierra Tucson made us do a meal in silence once. I laughed through the whole thing and had to leave the room. For some reason, I'm uncomfortable with silence. All of us just stare at one another and our food. We make funny faces at each other, attempting to make someone crack. Ruthanne glares at us, her eyes warning us of the consequences.

After breakfast, I smoke two cigarettes before taking the Druggy Buggy to the Beaumont Center. I take a seat in one of twenty chairs circling the group room. Posters of recovery art decorate the drab interior with clichés like, YOU ARE NOT ALONE! and EASY DOES IT.

This is my first group. I'm on level I, a beginner, and so I'm in a group for people in their first thirty days of treatment. We meet four times a week to focus on life stories and the first step, which discusses how we are powerless and our lives unmanageable due to our addictions.

Mounted on the wall of the group room is a big white board with the clients' names, the date they're admitted, therapist assignments, what level they're on, what assignments they are doing, and any consequences they may be experiencing at the moment. It reminds me of my stay at the hospital eating disorder unit, but the board there had the patients' names, nurse assignments, admission dates, and medications.

"Good morning, everyone," the therapist says. "Okay, let's begin." Like every group, we begin with the serenity prayer and a feelings check-in. A feelings check-in is where we go around the room, say our name, our addicted label, and choose a word to describe where we are emotionally. Today, I'm Greta, a bulimic, and I'm feeling happy and anxious.

"Okay, who has assignments to do?" the therapist asks. He studies the erasable board. "Oscar, you're up."

The circle of addicts' focus changes from the therapist to Oscar, who is reaching under his chair for his notebook.

"Okay. I'm Oscar, addict. And my assignment is 'life story'."

"Hi, Oscar," the room shouts back.

Oscar, an African-American thirty-something begins telling his life story of growing up in South Carolina. His father is a minister; mother is a nurse. He is the youngest of six children.

"I played baseball as a kid . . . I felt like there was a lot of pressure from my father to be perfect . . . so when I was twelve I started smoking marijuana . . . one year later, I was introduced to cocaine," Oscar says, showing his progression of addiction. "I was still able to keep my job, I was married, I could hide it well . . . I was going out every night, using prostitutes. Something I am ashamed of is that I gave a guy a blow job for drugs . . . ."

By the end of his story, it was clear to me how drugs ruined Oscar's dreams, relationships, and his life.

"Thank you, Oscar," the therapist says. "Okay, let's give Oscar feedback. Remember, speak in 'I' statements and share what you heard. Who wants to start?"

A couple of hands shoot upward immediately, as if they want to be the first to share. Others are more timid, looking around as if waiting for permission to raise their hands.

The comments are harsh:

"Oscar, what I heard was a lot of minimization. You need to get honest, Bro."

"You're still glamorizing the lifestyle and your addiction."

"I heard the unmanageability, but I didn't hear the powerlessness."

"What I heard, Oscar, was a lot of rationalization. A lot of 'yeah-but.' Anything after 'but' is bullshit. You have a lot of work to do."

"I would've liked to hear more about how it affected you and the consequences instead of just what happened."

"Thank you," Oscar says. He is only allowed to thank people for their feedback.

Oscar's in the hot seat. Tough crowd. I guess they want to tear you down and then rebuild from the ground up. I've spent a lifetime perfecting the skill of looking good in other people's eyes, and I definitely don't want to stop now. *I'm never going to be in the hot seat.*

"Oscar," the therapist starts. "I did hear a lot of minimization, denial, and rationalization. Recovery is about gut-level honesty, and I can hear that you have some stripping away before you reach nitty-gritty honesty. Keep working on it."

"Thanks for all your feedback," Oscar says to the group, emotionally depleted. Oscar stands up from his chair and goes up to the white board. He grabs a black marker and puts an *x* next to his assignment, signifying his completion. The room claps in support.

Later that day, we head to Hyde Park in South Tampa for another meeting. This time, though, it's an Overeaters Anonymous meeting, which I'm familiar with and looking forward to attending. The meeting is at the other Florida treatment center that Sierra Tucson referred me to.

We drive through a quaint neighborhood with street after street of historic bungalows. On Howard Street, the main drag on our way to the meeting, an eating disorder client Brie points out the "recovery Starbucks," where everyone goes before and after meetings. So much for anonymity I guess. I haven't had a Starbucks coffee in over a month, which is a record. "We go there sometimes before a meeting," Brie says. "Decaf, of course. And if you put half and half in your coffee it doesn't count as a fat." I don't care about that, I just want coffee.

We continue on our path to the meeting, heading on Bayshore Boulevard, the Beverly Hills of South Tampa. This area is beautiful. Tampa Bay on one side, mansions on the other. This is a place to live. We finally arrive, pulling into a Victorian-style house.

"Please sign in," a girl says behind the counter. "The meeting is upstairs."

The group of us walks upstairs and into the room. The eating disorders residents of Hyde Park are there. I feel like the skinny-minis are sizing me up. I have a sudden stomach and thigh awareness, which makes me feel fat. *If only you could've seen me three years ago. You would've seen that I can wither away, too.*

"Okay, let's get started. I'm Erin, I'm a food addict, and I'm the chairperson for this meeting."

"Hi Erin," the room shouts back. We do the requisite preamble and read each of the twelve steps aloud. "Would anyone like to share how you used a tool of recovery this week?"

"Hi, my name is Lindsey, anorexic, and I'd like to share on journaling. I recently began journaling again and it has really helped to get out my feelings. I can purge my feelings onto the page instead of in the toilet."

"Thanks, Lindsey."

"Hi, my name is Sarah, anorexic bulimic, and I will share on the food plan. The food plan really helps me because I don't have to obsess about what I will eat for each meal. Having a plan really allows me to have more time to think of things besides food."

"Thanks, Sarah," the room shouts once more. More hands go up.

"Hi, my name is Jen, and I'm anorexic. Reading the literature has helped me this week. If I just read out of the meditation books each morning, I start my day off better."

"Thanks, Jen."

"Hi, my name is Tracy, and I have been using prayer and

meditation a lot this week. I just keep trying to focus on my higher power's will for me, and not mine. I've been leaning on God when I'm struggling, and it has helped a lot.

"Thanks, Tracy."

"Hi, my name is Meredith, and I'd like to share on telephone and sponsorship. I've been speaking with my sponsor every day and calling women in recovery every day. It has made a big difference in my recovery because I don't do it alone. It's like the program says, 'It's a *we* program, not an *I* program.'"

"Is anyone visiting from out of town, here for the first time, or returning to Overeaters Anonymous?"

I raise my hand. "Hi, I'm Greta, and I'm a bulimic. I'm new to *this* meeting." I wanted to emphasize that while I'm new to Hyde Park, this isn't my first Overeaters Anonymous rodeo.

"Welcome, Greta," the room cackles back.

"Okay, while I'm passing around the *We Care* book, we can start open sharing." The *We Care* book is a binder filled with phone numbers of members so they may call each other for support after the group.

Hands raise around the room. A frail woman in her late thirties is chosen to share.

"Hi, my name is Cherie, and I'm bulimic and anorexic."

"Hi, Cherie," the room echoes.

"Hi everyone." Cherie smiles and holds herself in a meek way, as if she doesn't deserve to share. "Well, I'm just really struggling today. I've been in this disease since I was a young girl, and I'm really seeing some of the consequences. Today, I don't have any upper teeth because the bulimia and lack of nutrition has ruined them." Cherie is covering her mouth as she speaks in embarrassment. "I don't get my new teeth for three weeks. It's . . . it's just really hard. On the one hand, I want to quit the eating disorder. But it has such a hold on me. I know what I'm supposed to do, but I just don't do it."

I've never known a person to lose their teeth from bulimia. Thank God I'm receiving help now. Cherie is one of those chronic types that might not ever get out of her eating disorder. Prior to Sierra Tucson, I thought that would be me. But I'm changing my life. I'm learning it's possible to have a life without bulimia.

Accepting the Things I Can

've been here nearly a week and already feel at home. Unlike at Sierra Tucson, where everyone (not me) seemed to be "someone" or had a relative who was "someone," Turning Point has a bit more diversity, which I like. I loved Sierra Tucson, but it's nice to feel like I'm no longer at a boarding school for recovering addicts. Because I've now stepped down to a lower level of treatment, I have a greater sense of autonomy, which I like. I can't wait to get back to the city, but a small part of me wonders how I will survive outside of the parameters of the therapeutic cul-de-sac where I live.

Today, Friday, is a big day. We are going to the grocery store as a group. The eight eating-disordered patients pile into the Druggy Buggy. Ruthanne takes us to Publix Grocery Super Market. I'm wondering how this will work.

We walk through the automatic doors. I haven't been in a grocery store since my last binge. I'm buzzing with sensory stimuli. The cold temperature shocks my body and sends goose bumps racing up my arms. Seductive product labels and creative advertising jump out of every corner, trying to lure me to place their products into my cart.

I can see the bakery. The smell is decadent. My stomach flutters as I fantasize about the self-serve donut case. My mouth waters as we walk by without stealing a bite. I used to buy a dozen donuts and eat two or three in the store. I can still taste the sweetness. I can still feel the guilt, shame, and self-loathing that came shortly after I took the first sinful bite. I'm so happy I'm not in that emotional space anymore.

There will be no buying binge food today. I am in a group of eight and we are sharing two carts, which will only be filled with abstinent foods, monitored by Ruthanne. Today I'm not surveying which baked goods, candy, and frozen pizzas I should collect for my perfect binge. Instead, I'm one in a group deciding which brand of apples, frozen vegetables, and healthy grains to choose. The adrenaline rush of the grocery store is just as much of a jolt as riding the latest rides at an amusement park.

It takes us a long time to get through the aisles of the store. Each of us studies the labels of salad dressings and spends time deliberating over mundane details such as which flavor of couscous or turkey to buy. I see and feel people eyeing our group. At least an hour into our trip, Ruthanne lets us know we have to hurry up, which sends some into a mild panic. Ruthanne directs us to the cashier. She pays the bill for her eight food-finicky children. We make our way back to the Druggy Buggy with our two grocery carts full of food.

The other weekly outing is our Sunday night shopping trip to CVS for essentials like sugar-free gum and shampoo. It always feels more embarrassing when we pull into the parking lot of the drugstore than in the grocery store. CVS is smaller and people are in close proximity. I imagine as people look into the windows of the Druggy Buggy they are asking themselves why adults are not driving themselves to the drugstore. If they only knew.

We have fifteen minutes or so to buy what we want with our

escrow money. Once we pile back into the van, the house manager sifts through our plastic bags to make sure our gum is sugar-free and that we didn't buy anything to get high off of, like cold medicine or aerosol cans of whipped cream.

The highlight of every weekend for me is dancing on the asphalt in front of my house. I start with the warm-up I used to do in Michele's class in New York and progress to choreographed routines. I have to dance in socks because of the roughness of the asphalt, which blackens them and leaves huge holes. I've gone through at least six pairs, but it's worth it just to be able to feel dance in my life again.

I usually have an audience of at least a few recovery addicts watching me as they smoke at the picnic tables. Ever since I danced in front of the patients at Sierra Tucson, I have a new confidence and don't care if anyone watches me. I'm doing this for myself. I've always danced in front of the mirror. Now that I'm in the street, it's such a different experience. I'm not critically looking at myself for something to correct and improve. Moving in time with the song has a spiritual quality and makes me feel at peace. The sun beats on my face and warms my body, making beads of sweat form on my forehead. It feels good to sweat. Despite being absent from dance class, I still have my technique. My dancer's body is within me, too; it's lost just beneath my untoned flesh. I can deal with that. At least for today. I miss dancing. It's hard to think that my life as a professional dancer could be over. *Remember, Greta, God has a plan for you.*

I have the treatment center routine down, as we have the same groups every week. This morning I have Big Group, which is where we confront each other on behaviors we observe going on in the therapeutic community. I sit down next to Teresa. Confrontations are my nightmare. I've tried to avoid them my entire life.

"All right everybody, let's begin." Mike is the therapist for this meeting. He's about sixty years old and doesn't put up with addicts' bullshit. Everyone holds hands, and says the serenity prayer. "God, grant me the serenity to accept the things I cannot change, the courage to change the things I can, and the wisdom to know the difference." At least something is familiar.

The therapist scans the room and invites us to check in about our feelings. One by one, each client shares their diagnostic label and their feelings: grateful and empowered, angry, motivated, numb, confused. My insides quiver as my turn comes closer. I hate speaking in groups of people I don't know.

"Hi, my name is Greta, I'm bulimic, and I'm feeling grateful today." In recovery, saying you are grateful is sometimes equivalent to how normal people say they are fine when asked how they are. If you don't feel like really digging into your feelings, grateful always sounds good. Today, though, I'm truly grateful. For once, I'm not throwing up.

Suddenly, Joan, the senior house manager, brings in a fancy purse and places it in the middle of the floor. Everyone just stares at it wondering what the point is. One girl, Esther, looks guilty. Turns out, it's her purse in the circle.

"Does anyone want to discuss this?" Joan says, opening the purse and dumping out the contents. The purse is filled with cereal that Esther hid.

"It's mine," Esther confesses.

"Esther, why is the cereal in your purse?" Mike asks. It turns out Esther was hoarding her cereal and eating it late at night.

"I know I should've asked for help, but I just didn't. I'm sorry to everyone for lying."

"Esther, we are going to have you go to 4A tonight and stay in the house manager's building. It's not a punishment. It's to keep you safe," Mike says.

Anytime a client has a consequence, the community is reminded that it isn't a punishment or because we're bad. One part of recovery is being accountable for one's actions. Something I wasn't too good at when in the grips of active bulimia.

The professionals at Turning Point don't coddle their clients. That isn't their job. They are here to help us face the truth about our lives and our addictions, and help us learn how to live "life on life's terms," as stated in the *Big Book*. That means facing and accepting what life brings, without using maladaptive coping skills and self-destruction. Today, I'm facing life head on, wearing the protective armor of coping skills I've learned in treatment, such as reaching out for help and positive self-talk. I remind myself daily of the phrase, "God doesn't give us more than we can handle." Recovery clichés really are quite helpful.

While the majority of the therapeutic community's time is spent trying to get better, it doesn't mean we're angels. Hardly. Addicts try to get away with anything they can, especially when they don't have their precious substance to hold onto for security.

One evening, while I'm smoking outside our house, Pauline backs her Mercedes into the driveway.

"Hey, come over here," she whispers as she opens up her trunk.

"Here, take this bag and bring it into the kitchen." Pauline is on level II, which means she can drive herself to meetings. I look in the bag and see the contraband. *Hallelujah, it's coffeeeee!*

"Where do you want me to put it?" I ask.

"Get the decaf coffee can out of the cabinet." I follow her instructions.

Pauline comes back inside and pours the regular coffee into the decaf Folger's metal tin. Teresa and Andrea have joined us in the kitchen. We huddle around the coffee as if we're admiring Tiffany diamonds. I haven't had coffee in weeks. *Caffeine isn't my problem; purging is.* I'm not shameful for my involvement in our coffee scheme.

The next morning at breakfast, our eyes sparkle as we sip our regular coffee and exchange glances of triumph. House manager Ruthann doesn't have a clue.

"OKAY," MARGE, THE therapist, begins. "I want you to take out a piece of paper and write down ten thoughts you tell yourself that influence your eating disorder behaviors."

Part of therapy is doing assignments. A lot of clients hate them, but for me, it's a chance to grab my gold star, to do them perfectly. At heart, I know that I need to change my perfectionism. It's not that I even want to be perfect. I just feel so imperfect compared with my peers that I have to overcompensate to make up for my inadequacies.

We do as we are told (the eating disordered population are mostly approval-seeking perfectionists). I'm a pro at this exercise. I'm constantly speaking critically to myself and feeding myself negative messages.

I volunteer to read mine first. Everyone is focused on me. I'm jittery inside, unsure as to how I will sound. I have felt these jitters since first grade, whenever I raise my hand to answer a question. Even when I think I have the right answers, or even when there are no correct answers, I'm sure I will somehow mess it up.

My List of Negative Self-Talk

1. *If I eat this, I will get fat and no one will love me.*
2. *I'm only attractive when I'm really thin.*
3. *The only way you will get love and attention from your family is if you stay sick.*
4. *You deserve to punish yourself because you're a horrible person.*
5. *You have a few hours before anyone comes home, so you can binge and purge.*

6. *Everything has gone wrong today, so I might as well fuck the rest of my day by bingeing and purging.*
7. *I don't deserve to eat today.*
8. *If you don't eat, you will be stronger, and people will love you.*
9. *If you're fat that means you are lazy, ugly, you won't get married, and you will not be successful in life. Therefore, you cannot keep this food in you.*
10. *You are a fat piece of shit and you need to run more miles on the treadmill.*

"How'd that feel to read?" Marge asks.

"It didn't feel great," I say, my eyes now on the floor.

"Can you give me a feeling word?" My eyes are welling up with tears.

"I . . . I feel sad." Now the waterworks are flowing. "I mean, I wouldn't talk to anyone else that way. It's just so cruel."

"You're right," Marge says. "It is cruel. Group, let's give Greta some feedback."

"Greta, I can really identify with your list," Nancy says. "I have a lot of the same things on my list."

"You're really hard on yourself," Teresa says. "You deserve to eat and be happy." Teresa's comment has triggered a burst of positive energy.

"Yeah, Greta," Pauline says. "You are so much more than a body. I've only known you for a short time, but you're funny, pretty, and really kindhearted."

It's hard to take in the positive affirmations from the group. I feel ashamed and embarrassed for my thoughts. I don't want to think negatively about myself, but I can't seem to change it. I know what they are saying. I've said these same words to myself before hoping to believe them. The truth is that my thoughts have nothing to do with my body. They represent how inadequate and insignificant I feel on the inside.

After everyone reads the lists, the hardest part begins. We have to take each negative internal message and counter it with a positive message. The group becomes sullen, as the reality of our low self-esteem punches us in the gut. It's so much easier to verbally abuse ourselves.

I do my list, but I don't believe what I'm writing. I want to believe it, but I just don't. This is one of those "act as if" exercises. Maybe I don't believe it today, but if I keep telling myself these positive messages, I'll brainwash myself into thinking things like, "It's my inner beauty that counts!" and "I deserve to eat as much as everyone else!" Suddenly, I'm feeling fat. I know fat is not a feeling, but for me it is.

ABOUT HALFWAY THROUGH our next big group meeting, Joan makes another appearance and comes in with a bag. Here we go. Esther has put cereal in her purse again. *Back to 4A you go, girl.* My stomach turns and my conscience freezes with guilt when I see the contents inside of the bag. It is the Folger's coffee tin. We've been caught!

"Alright," group leader Mike says with a stern face. "Who put regular coffee in the decaf tin?" The guilty culprits in the room, which is more than half of the eating disorder clients look at one another. I maintain an heir of innocence. *Coffee? What in the world are you talking about? I thought I was drinking decaf.*

Confusion has struck the room. The clients have been in the dark. "You mean someone has been drinking regular coffee?" Joe says.

Maggie raises her hand. "I wish someone would just admit it. It isn't fair to the rest of us."

Steve chimes in. "This is about being accountable. Own up, people." That was the clincher. Pauline, Andrea, Teresa, Lisa, Tricia, and I were not innocent. As I had learned, recovery is about taking responsibility. The ninth step of all twelve-step programs says, "Made amends to all people we had harmed, and promptly admitted it." I was only on step three. Do I really have to go there, yet?

Pauline, the instigator of the coffee scheme, could no longer take it. "I . . . I did it. It was me. I brought the coffee in. I brought it in from the trunk of my car." Her chin drops, her eyes plummet to the floor. Shame has taken over her body. The room is awestruck. Hands raise. It's time for confrontations.

"I am really upset that you would bring coffee into the community. It isn't allowed, plus you were being so sneaky. It makes me angry and I don't feel safe in the eating disorders house," Maggie says.

"Yes, it isn't safe for the community. This is a therapeutic environment. I feel angry and scared," Tracy says.

"When are you going to get serious, Pauline?" Pauline now has been reduced to tears. My ears are on fire. I'm sure by this point my face is radiating a nice shade of cherry, enflamed with guilt.

Suddenly, these people who shoot dope into their veins, who drink enough alcohol to cause liver failure, and who digest enough Twinkies to send a diabetic into a coma are afraid of a coffee tin. *Get off of your recovery high horses, people.* In an instant, the room has turned from food and drug junkies to recovery saints. *Lord, help us.*

"Why did you do it, Pauline?" Mike asks.

"I don't know. I just wanted coffee."

"Get honest, Pauline. Why did you do it?" The group is looking around, half confused as to why Mike is asking her again. We don't know what the ulterior motive was.

"Well, I wanted to get away with something. I guess I was rebelling. And it was a way for me to fit in."

"Now that's honesty," Mike says. He looks around the room. "So, I know you brought in the coffee. Who else is involved?" Mike looks around the room.

One by one, we raise our hands. I don't want to look at Mike. Mike is my favorite therapist and I don't want to disappoint him. I have been doing so well, completely focused on my recovery. That coffee scheme was during my first month; however, we have continued to drink it, as Pauline has bought regular coffee every week.

I had so many other issues to think about than this coffee deal. Yes, I know it's not allowed in the therapeutic community, but at the end of the day, it's just coffee. Yes, the professionals say that coffee is a drug, coffee is a mood changer. *Whatever. I will always drink coffee.* I had to start really thinking about whether I was going back to New York.

After group, I go up to Mike. "Sorry about the coffee, Mike."

"What were you thinking, kid?" Mike asks.

"I don't know . . . I just wanted regular coffee," I said.

"You just wanted to get away with something," Mike said.

"I suppose," I say.

"You're smarter than that. Don't try to cheat the system. Recovery is about being honest."

"I know." As a consequence, I have to wash the Druggy Buggy this weekend. I've got to be more serious and stop fucking around.

I desperately want to go back to New York. I am contracted to start Rockettes again in a month. I haven't signed my contract yet, and I certainly am not in shape for dancing. Even though I weigh six pounds less than when I started treatment, I cannot envision squeezing myself into my green velvet costume. Just thinking about it makes me want to purge. When I came to Tampa, I was sure I would leave as soon as my ninety days were up. But now I'm not so sure.

A few days after big group, I am scheduled for a consultation to talk with Mike about my future. I plop myself down on his fluffy couch. I've never been in his office since he's not my therapist. I look around, noticing all of the pictures. I see he's married, and he has at least one child, a daughter. He's a grandfather. Even though his icy blue eyes can scare an addict into sobriety, I imagine he's as soft as a Ho-Hos cream-filled cake when it comes to his grandchildren.

"I just don't know what I should do, Mike," I say.

"What do you want to do?" Mike asks.

"Well, I want to go back to New York. But I just keep thinking of what I hear in the program: Anything you put before recovery you lose. I lost so much before. I know that there is a lot of Overeaters Anonymous support in New York, but there is also food on every corner. Plus, I can't really imagine dancing at this point.

"Do you want to dance?" Mike says. That was the question.

"Of course I want to dance. I love taking classes. That's what I miss. I don't miss performing, though. I don't even want to go back to the Rockettes. I'm too fat."

"Kid, I hear all about what you *want* to do. But what do you *need* to do?" I knew what I needed to do.

"I . . . I need to stay here, and continue my recovery. Plus, I feel like I have some trauma issues to work through." I had my answer.

I know I need to call my mom and dad to let them know my plans. My sixty days are about up. Now that I'm going to stay in Tampa, I want to see if I can have some more time in treatment. I feel like the extra month can help me. I haven't even touched my sexuality issues. Plus, I need to figure out where I'm going to live and what I'll be doing with my life in Tampa. Eventually I know I will go back to school to become a therapist, but I need a significant amount of time in recovery before I can deal with that level of stress.

I call my father first, since he's paying for this leg of treatment. I'm nervous. I've asked for so much money from him in the past and now I have to ask again. I feel like such a burden.

"I think I've decided to stay in Tampa for a while, Daddy. I think it's too soon to go back to the city."

"Oh really? You think you want to stay?"

"Well, I want to go back, but it's probably better for me to stay here. I wanted to know if I can stay one more month in treatment, and if you'd help me with a car?" I tried to say it fast, so it wouldn't shock him.

A short silence ensues. It feels like an eternity.

"Well, I can either buy you a car or pay for another month of treatment, not both." My first instinct is to take the car. I have to have a car in Tampa. But I know another month at Turning Point will benefit me.

"Okay, well let me think about it, and I'll call you back tomorrow. Thanks, Daddy."

We hang up, and I consult my house members. Pauline tells me that I could probably lease a car for little to no money down. Pauline doesn't know my credit score. If I did that, my mom will have to cosign. In her efforts to help in the past, she's cosigned on so much of my bullshit. I feel sure she'll cosign for me now when I'm doing well.

I am nervous to call my mom. She is always talking about my dancing. It's the one interesting topic of substance we share.

"Hi, Mommy," I begin. "Well, I don't think I will go back to New York just yet."

She pauses before answering. "Oh, you don't?" Her tone is one of curiosity, unlike the criticalness I assumed she'd have.

"No, the treatment team thinks it would be better for my recovery to stay here for another month to work further on my issues. I'm in this trauma group, and I think it would be a good idea if I stayed."

"Trauma? Did something happen to you?"

"Well . . . yes. Besides, I have other issues to work out still." *My sexuality for starters.* I hope she doesn't ask what trauma I have. It's hard enough for me to think about—let alone talk about—being fondled as a child or being sexually taken advantage of as a teen-ager by a boy in my basement, with my mom and Kent just one floor above me upstairs. I had wanted to yell out for their help, but felt shame for putting myself in that position. Other traumas that I've experienced, like the divorce, the car wreck, and the Northridge Earthquake, I can easily talk about. But the deepest ones are still too tender and difficult to say aloud.

"Another month? I know you have issues, but you aren't on a vacation, Greta." Wow. A vacation. Yes, clearly that's what I'm on. Had she forgotten how exhausted she was after doing just five hours of therapy at Sierra Tucson?

"You really think that I think I'm on vacation?"

"No. But, Greta, at some point you have to get back to the real world. Get back to work. Get back to doing normal things, like a normal person."

There was that word. Can't she see by now that I am not normal? Not too long ago, I was puking thirty times a day, cutting myself, and wanting to slice my wrists open. I was trying everything to escape my reality. Now is actually the first time in my life I am living in the real world. The first time that I'm dealing with the reality of who I am.

"Have you told your father yet?"

"Yes. Daddy's paying for it. He said I could get a car or another month of treatment."

"Wow, that was nice of him. Well, okay. Whatever you think is best." Before we hang up, I ask her if she will cosign on a car for me if I choose to stay in treatment another month. She agrees. She always comes through for me in this way. After learning the word "enable" at Sierra Tucson, she's much more comfortable cosigning on a car than on my usual bulimic bullshit.

We hang up. *Vacation. The nerve she has.* One thing I've learned in treatment: I have a hard time letting things go. A recovery cliché zips through my mind. *Resentment is the number one offender. You will relapse over your resentments.*

The next day, I call Radio City to tell them I won't be returning. It isn't a problem. There are plenty of girls waiting to audition to take my spot.

On Tuesday, I walk into grief group expecting a typical meeting. I'm surprised to find candles lit, the lighting dimmed, and a large casket in the front of the room. Instead of a circle of chairs, seats are lined up in rows just like they are at a funeral.

"Today, we are having a funeral," the therapist says. "It is your funeral she says to the room. Imagine that you have died from your addiction. We are going to write a eulogy about who you were. You can also write about your disease and how it brought you to your death."

I feel sad already. I write the eulogy, looking up from my paper periodically to imagine I'm lying in the casket. I think about my family, friends, me. I think about all that I've been willing to give up for this stupid eating disorder.

A fire grows inside of me. I'm not going to let the eating disorder win, damn it. I don't want to die. I want to live! Maybe that was the thing I needed to shock me into a life of health. I hope I never relapse again.

FOR MY LAST month, I decide to focus my energy on my sexuality issues. I feel like a disgusting person. I have homophobic messages ingrained in my mind. I'm sure that no one in my family will accept me if I'm a lesbian. Even in drunken hazes of make-out sessions with women, my inner shame disables me from truly being free in the moment. It's hard to be free or enjoy myself when the common gay discriminatory remarks I grew up with whisper in my head . . . *homos, faggots, strange people.* It's a real libido booster, let me tell you.

"Greta," Pauline says one afternoon. "At some point, you have to start living for yourself. That's what I have to do, too." Pauline hasn't come out to her parents yet and she's almost forty.

"I know, I know. I just can't imagine having to tell my family. My dad will probably disown me."

"And if he does . . . well then that's his loss."

"Yeah, I guess." It feels much more like it would be my loss though.

What I can't get over is my own internalized homophobia. I don't want to be this way. My sexuality isn't some chosen lifestyle. I have felt strange and different my entire life. I couldn't ever figure out why I thought it was gross when my friends started having intercourse with their boyfriends, why I had no interest in boys, or why I couldn't picture myself walking down the aisle in a white gown, getting ready to marry a man.

I never fantasized about the fairytale wedding and living the white picket-fence dream. Instead, I always had a strange and fearful vision that I would be wrongly accused of a crime I didn't commit, and be sentenced to a life in prison. But that's another story. In a sense, though, that fear has played out. I've been imprisoned by bulimia and my internalized homophobia for nearly half of my life.

As much as I want to work on the sexuality issue, I'm too ashamed to talk about it with any of my therapists. I can't discuss it because I'm not really ready to fully admit it to myself. Deep down, I know this is my truth. I. Am. Gay. I'm ashamed to say the words, even in the privacy of my own mind. One thing I do know is that if I don't come to terms with my sexuality, I will relapse. The label of bulimic is much more socially acceptable than the label of homosexual. I think I will just have to deal with this issue at a later date.

SINCE I'VE BEEN practicing recovery for several months now, I'm shocked one night after Meta when I have the urge to overeat.

I don't know why. I finish my measured food exchanges. I have had enough. As soon as I finish my last bite of cottage cheese, peach, and Kashi concoction, I feel an immense sadness that the meal is over. I go to bed confident the urge will go away. Intuitively I know that this means trouble. I've been floating on the pink cloud for way too long.

About two hours later, after my roommates are asleep, I slither out to the kitchen. I open the cabinet, grab a bowl, and pour myself some Kashi with skim milk. I never eat cereal with milk, especially skim, unless I am going to binge.

The slogans of the program and the tools I should be using poke and prod my eating disorder mind, but to no avail. I know I should wake up a roommate, go talk to a house manager in 4A, or call someone from the program. But I don't. I want that cereal.

The next morning, no one knows that I ate the cereal. I feel guilty. What is happening? I'm on a pink cloud. I'm not supposed to want to eat extra food. *But it's natural, Greta. You're a food addict.* My head was so clouded by images of Kashi that I completely forgot that I have to get weighed today. *Stupid, fat-ass!!* I'm depressed. More than depressed, I'm scared. I know if I don't use the tools of recovery in treatment, then I certainly won't use them outside of treatment.

Later that day, I see the dietician and confess my extra bowl of cereal. She's compassionate about the slip and suggests I speak with a therapist. I talk with Mike about it. He's firm, but empathetic. We review my motives and he asks me how I could've used healthier coping tools.

"You know what you need to do, Greta. You need to keep practicing using the tools of the program." He's right. I did know. I'm trying to tell myself that this was just a minor slip. After all, I only ate a couple of extra meal exchanges for the day. But at my core, I know the slip has huge implications. Not because of what I ate. It's huge because of what it signifies, which freaks me out. Not only does it mean that I still eat out of emotional hunger; more importantly, I didn't use any of the coping skills I've learned. I didn't ask for help, even though I had an entire therapeutic community, and I ate regardless of the emotional consequences. The scariest part is knowing that I might purge again, my ultimate nightmare. If I can have a slip of overeating, then I can certainly have a relapse with bingeing and purging. For so long, I've thought that bingeing and purging is out of the question. I am reminded that I have to be diligent in my recovery.

With only a couple of weeks left at Turning Point, I have to find somewhere to live. Teresa and I want to live together, but Mike says that two people new in recovery will relapse. Darn it. The other option is a halfway house. Ick. One of the eating disorder clients who was getting out at the same time I was said that she checked with her mom, who's a social worker, and I can live with her. I'm excited.

I used to tell Heather, my therapist in New York, that once I stopped dancing I would pursue my degree in social work to fast track my way to becoming a therapist. Not only do I have a place to live, but also the mom I'll be living with is a social worker at the school where I can pursue my undergraduate degree. This is a sign from my Higher Power. I'm beginning to think that God has a blueprint of my life already made up for me.

Pauline takes me to a Honda dealership to see if I can get a deal on a car. To my benefit, and with the help of my mother cosigning, I'm able to lease a car with no money down. I not only have a place to stay, but also a car. *See, Greta. Higher Power works in mysterious ways. You just have to ask for help and take the next right action.* I make a promise to myself that this car will not become a fast-food binge pit, filled with bags and wrappers up to the seat, like my other cars.

I CAN'T BELIEVE I'm down to my last day of treatment. It's hard to believe that I will be leaving treatment soon. My emotions vacillate between terror and excitement. What if I can't do this? What if I relapse? The therapists have said that most of us will relapse. That's just how addiction recovery goes. Maybe I will be the one who doesn't. I'm keeping my fingers crossed. Regardless, I'm excited to create a new life here. It feels like the possibilities are endless.

My mind flips back in time, seeing the snapshot stages of my recovery. Even though I had the recent slip with the cereal, I cannot imagine purging again. Although I know it's possible. I can't believe I binged and purged for nearly half of my life. I can't believe I had so much self-hatred. I used to shed so many tears over my inability to change. I cry now because I am sad to think of what I put myself through. I spent so much of my life being physically present, but absent emotionally, enslaved by the eating disorder.

My previous therapists used to describe me as extremely intuitive and insightful because of my level of self-awareness. It was great that I knew the pathology of my perfectionism, my needing to be a people pleaser, my need to stay sick with the eating disorder so that I would receive a shred of affection when I was in crisis. The fact is, though, that knowing the myriad insights I'd learned in therapy didn't help me to change my behavior. As the twelve-step program has taught me, "Self-knowledge avails us nothing."

Since being in recovery, I've slowly started to change what I was never able to do on my own. All of the therapy groups, the recovery slogans, the assignments, reading affirmations and daily meditations, and having the willingness to practice saying embarrassing things to myself in the mirror like, I love you, has helped me. In order to believe that I don't need to be perfect, I've had to challenge all of my thoughts and beliefs, and then practice not being quite as rigid in my thinking. This means that I've had to stop assigning values like "good" and "bad" to my behaviors and myself. It plays out in my food, too. I'm practicing not having to eat in a certain order. I'm learning to not have to save the best bite for last, which, as embarrassing as it is to admit, scares the shit out of me.

I'm learning that I can't please everyone all of the time, and that I have to accept it. Now when I displease someone, I do a lot of self-talk and remind myself that I'm okay even if someone might be upset with me. It doesn't make me bad, which used to translate into *I'm fat and have to lose weight.* This all can sound so rudimentary. Intellectually, I've known these realizations for years. But to actually believe the truth means that I show my truth in my behavior, and I haven't been able to start doing that until now.

My peers have given me so much support. At Sierra Tucson, the therapists used to have us decorate and write in bright-colored magic markers notecards with positive affirmations and qualities we like about our peers. I've kept so many:

- *Greta, you're so sincere and nice.*
- *Greta, you're such a beautiful person, inside and out.*
- *Greta, you're caring and such a great listener.*

My first twenty-eight days in treatment, I read these blanket affirmations, but was unable to digest them as truths for myself. After 120 days, I'm a bit closer.

We do positive affirmations in Tampa, too, but we also do peer reviews and tell each other what we observe that they need to work on. As much as I don't like to hear what I need to change, it has been most helpful.

- *"Greta, I'd like to see you show more anger."*

- *"Greta, I'd like to see you stop smoking."*
- *"Greta, I'd like you to stop trying to present that you're okay."*
- *"Greta, I'd like you to stop worrying about what others think."*

In order to change, I've had to look at all parts of myself, not just the eating disorder behaviors.

Thinking about my life without dance or the eating disorder —since my identity has been knotted up in both—is probably the most challenging part of this process. Deciding not to go back to New York and dance has been the toughest decision. I have been known as a dancer since elementary school. Dance meant more to me than just learning and performing steps. Dance was how I knew I was good enough as a person. And on most days, I never felt good enough.

Until now.

Treatment and my ability to be in recovery provide a level of confidence that I've never before had. It's the truth. I've lived without the eating disorder for nearly four months, and have felt sustained happiness for the first time. I feel happier now then when I've been at my thinnest (which isn't saying too much since extreme thinness actually feels like shit). I'm learning that I can have an identity outside of dance and an eating disorder. Perhaps the greatest realization of all is that I am actually successful just being me, without having to do anything.

It's officially my last day in treatment. We are doing what's called a "coin-out." I'm presented with a bronzed coin, which has the Serenity Prayer engraved on the back and a "One Day at a Time" on the front. The coin is passed around the circle, and as the coin passes through each person's hands, they put drops of hope into my coin for me to "take with me" on my recovery journey. I hate being the center of attention and having people say really nice things to me, but I have no choice today. One by one, my recovering peers give me their well wishes. Into my coin they give me drops of strength, courage, serenity, self-love, acceptance, self-care, motivation, continued abstinence, support, and peace—all things I will need each day to continue to recover.

"Thank you, everyone. You all have helped me so much. I will never forget this time in my life. It's not goodbye. I will see you at meetings. I will miss you all so much," I say.

At the end of the group, I hug everyone goodbye. I give Mike, Lori, Ruthanne, and Marge a big hug, promising to visit them regularly. I smoke one last cigarette with my friends before taking off. I get in my new Honda and start to drive away. In my rearview mirror, I see all of my friends hanging back. A tinge of sadness comes over me knowing that I'll never again have the security of the therapeutic community. I have the tools to cope with life without abusing food, and I feel ready. Driving off into my new life of recovery, I don't know what all of the pieces look like or how I fit. But I do know that I'll be able to handle whatever life throws my way. I have a new level of faith and I know that I will be okay.

Yes, I am still Greta, recovering bulimic at twelve-step meetings. That will not change, as I will always be in recovery. With that acknowledgment comes a level of responsibility to take care of myself daily. I'm okay with that. Using recovery tools, coping skills, and following a meal plan—things that seem tedious to others—provides me a level of freedom that I never had when I was bingeing, purging, or starving myself. Today I can say I am free, and for that, I am truly grateful.

Finale

A lot has happened in the ten years since I went into treatment. In recovery, I've spent a lot of time talking to women about food addiction at a particular Starbucks in Tampa that is frequented by people in a variety of twelve-step programs. At these coffee outings, so many women struggling with eating disorders have asked me, "So, how did you finally stop throwing up?" It doesn't matter who's asking. Every time, the same pair of inquisitive eyes stares at me waiting to hear the magical formula. I recognize the look, a hybrid of desperation and curiosity. I sat in their seats years earlier at the exact same outdoor café, sipping my eight-Equal-skinny-sugar-free-cinnamon-dolce-latte, eager to know the answers to my own food problems.

My reply to these struggling women used to be the standard "program" responses—I prayed for the willingness, I followed a food plan, I attended meetings, I used a sponsor, I picked up the telephone, and I read the *Big Book*. What I omitted from my practically rehearsed recovery speech is that I did all of those things for years, prior to my last treatment stint, and I still threw up. The episodes may not have lasted as long, and I may not have exchanged my job for a binge/purge session. My life may have been manageable, for the most part; but the disordered eating was still present, which, more than anything, was frustrating. It's like an ingrown toenail that you dig out, only to feel it a couple of months later poking at a tender corner of your big toe as you walk the streets of Soho in your Tory Burch flats. You thought you'd taken the necessary preventative measures, but no, there it is again. I didn't understand why I couldn't stop the behavior entirely, especially when it caused me so much angst.

The fact of the matter is that few recovering food addicts skip that last meal or purge once more and decide, "That's it. I've had enough. I never want to do that again!" To date, I've known only one person, out of hundreds, who has had "perfect" recovery—a hard lesson to swallow for the rest of us eating-disordered perfectionists.

Now when I talk to women, my reply is the truth. The truth, for me, is that while all of the tools of the program helped prepare me for and maintain the eventual abstinence I have—thwarting many would-be relapses along the way—I only stopped throwing up once I had something better to do than throw up. In essence, I stopped once my life was full enough that I no longer needed to fill up on food.

It wasn't enough for me to "pray for the willingness to be willing." I could attend meetings all day long, and I could serve myself the measured portions of protein, carbohydrates, and fat listed on my food plan. However, until I not only found a better replacement but also believed I was worth the recovery, I still found myself ordering an extra-value meal at the McDonald's drive-thru just moments after attending an Overeaters Anonymous meeting! Eating disorders are just that sneaky.

As a person who has spent a lifetime keeping up appearances, attempting to have people think that my life is perfect, at all times,

I'd love to describe my recovery as this beautifully wrapped present with tidy creases, perfect edges, and shiny bows. While my recovery has been the greatest gift in my life thus far, my process has been more like a helix of harm-reduction.

As to how I finally believed I was worthy of recovery, and that there was a replacement better than food to fill my needs, that's definitely the harder question to answer. Contrary to the belief that anorexia and bulimia are vanity disorders—*So, why don't you just eat three square meals a day and get over yourself*—eating disorders are layered with emotional complexities. Throughout this process, my behavior has paralleled just how resolved, or unresolved, my emotional issues were. My abstinence didn't stick until I began to unravel all of the knots of mixed messages I'd received or misconstrued, along with the irrational and negative beliefs I'd chosen to buy into. Once unraveled, I began to string together new beliefs about myself. Eventually, I wove together a strong sense of self that consisted of more than me being a dancer or having an eating disorder.

After chipping away at the bulk of my issues—the constant sense of inadequacy, the perfectionism, the people-pleasing, the shame of being a lesbian growing up with a homophobic father, feeling like I had to overcompensate just to keep up with my peers—I had to learn to accept who I was at the core of those layers. I also had to accept that by feeding my eating disorder, I'd chosen to forfeit my dance career, as well as nearly two decades of time. Until recovery, my most vivid life memories are not of experiences with family and friends, nor the thirty-two countries I'd traveled to while dancing, but rather the food I'd eaten and where I'd puked.

Even after I resolved those issues, I still hadn't healed my biggest one—the gnawing ache of emptiness. I felt unlovable. And because I could never get the affection I needed, I constantly craved it. As it turns out, I'm not unlovable. My parents just aren't the type that hugged or regularly said, "I love you." I was so angry because that unfulfilled need was the greatest source of my emptiness.

At the end of the day, though, I had to accept my parents for who they are—wonderful people with their own set of limitations, who'd do anything for me. It's not their fault that they didn't show affection the way I would've liked them to. Between their families of origin and the era they grew up in, talking about issues and showing affection just wasn't something they did.

My parents have done so many wonderful things to show they love me. They did so much when I was in the eating disorder, like constantly bailing me out when I had no money and paying for treatment. They've allowed me to have every opportunity I've wanted and have given me anything materially I've ever asked for. My mom spent years sewing dance costumes, cheerleading outfits, correcting and redoing school projects when I wasn't equipped, and so much more. I had to stop looking outside of myself for nurturance. I had to learn how to nurture myself.

Today, my life is amazing. I've been in a relationship for more than eight years with a wonderful woman, Lisa, who loves me just as I am, no matter how I look or what I'm achieving. She makes me feel beautiful in every way, and her love fulfills me more than I imagined a relationship ever could. She honors all of my feelings and isn't afraid to tackle my negative or irrational emotions. She's my wife, best friend, partner, and lover. I thank God that I have her in my life.

My mother and Kent have welcomed Lisa into our family not only because they like her but also because they know how much Lisa means to me. I still wonder if it's awkward when my mother has to decide whether to tell her friends or coworkers that I'm gay, and if and how she responds when they ask if I'm married. Thanks to Massachusetts's law, yes we are! My mom and I don't discuss it. Everyone acknowledges I'm a lesbian—mainly because of the pretty brunette

I keep carting around to holiday functions, so there is no denying it—but we don't openly discuss my gayness.

Because my father is a conservative Republican, he still isn't too keen on me being a lesbian. Last Christmas, however, he was able to crack a joke or two with Lisa, and he even held a political debate with her, all of which says he's experiencing his own level of acceptance. But on that trip for the holidays, he told me, "I call myself a 'homo' whenever I make a bad shot on the golf course." Something in that sentence tells me there's a disconnect somewhere. Call me crazy, but I don't think he's pieced together or admitted to himself that his daughter and the "homo" he refers to are essentially one in the same. Recently, though, he came to visit me and the three of us went to Lombardi's for pizza and he was able to make conversation with Lisa, laugh, and have a great time. Progress, not perfection, I guess.

After living and recovering in Tampa for eight years, I, along with Lisa, am living in New York City again. Lisa and I spend most of our time with our two golden retrievers, Jack and Emma, and are planning to have a baby in the near future.

I graduated with my master's in social work from New York University's Silver School of Social Work program. I finally am fulfilling my dream of helping others; I work as a therapist at an eating disorders clinic and have a private practice in New York City. Working as an eating disorder therapist not only fulfills my dream, it also helps my recovery, reminding that I never want to struggle in that way again.

I feel like I'm in a constant growth spurt, which, of course, comes with its own set of growing pains. I'm certainly not the chameleon I once was with my defining qualities being my achievements and the numbers on the scale. I have thoughts and opinions of my own, and if you don't agree with them, I still have a right to think, say, feel, and believe them for myself.

Recovery has been about getting to know myself. Although my

sense of self is stronger than ever before, I'm amazed at how much more I have to learn. Karen Carpenter said it best, "We've only just begun."

The one thing I've learned about eating disorder recovery is that no one solution is right for everyone. Some people need Overeaters Anonymous meetings, while others can do well with just therapy. From being in contact with hundreds of women with eating disorders, I know that some people will fully recover from an eating disorder, never having to obsess about food and weight again. Others, like me, will always be in recovery. I used to think that every person with an eating disorder was a food addict like me. That simply is not true.

While, for myself, I think of food as an addiction, I do not have an allergy to sugar that many in Overeaters Anonymous endorse. I regularly eat it in moderate amounts. I do my best to listen to what my body wants. Although I don't attend Overeaters Anonymous meetings anymore, I still use the tools of the program daily. Thankfully, they will be forever ingrained. One of the most important aspects of all twelve-step programs is the slogan, "It's a 'we' not an 'I' program." Even though I don't go to Overeaters Anonymous anymore, I have many supports in my life. Continuing to ask for help and not suffer alone is key.

My relationship with food isn't perfect. We aren't lovers, but we also aren't enemies. Based on my weight and the food choices I make, I appear very "normal," according to Western culture's standards. In fact, I've had some recovery "friends" judge me because I eat like I've never had an eating disorder. I've moved away from treatment center food plans, weighing and measuring, the no-sugar and no-flour abstinence plans, and eating out of Tupperware containers. I don't have "trigger" foods, anymore. Every food is safe, as long as I'm in the correct emotional state. With that said, I know that if I don't use daily tools and coping skills, I would eventually binge and purge.

The best part about my life today is that I can be present. I don't miss out on life because I'm privately scheming when I can excuse myself from the dinner table to purge. I don't have to miss out on social events because I feel too fat to attend. I'm able to listen to whoever is speaking to me, instead of obsessing about how my skin is resting on my stomach when I sit, and whether they notice. I don't berate myself when I see myself in a window storefront.

There aren't any more secret trips to the emergency room because I have chest pains. I no longer have constant sore throats and ear infections, nor do I bump into walls due to electrolyte imbalance and dizziness, or have bruises after taking a bath because I don't have enough fat to protect myself from the hard porcelain tub. I don't steal my parents' credit card numbers, nor do I pawn my jewelry in exchange for binge food money. My parents trust me today. When I tell them I have to "run errands," they believe me, and that's what I do. In recovery, I have learned to live honestly and responsibly, and I do this to the best of my ability, one day at a time.

Perhaps the greatest joy in my life is that I don't wake up with that sense of dread, wondering if I'll make it through a day without puking. The shame and self-loathing that accompanies relapse no longer rents space in my head. I don't have to wait for tomorrow; today is a good day. I no longer traipse the streets of New York City in search of the perfect food that will fill me up in just the right way. Nor do I know where the best private bathroom stalls are in Manhattan (luckily, I've forgotten).

I'd be lying if I said I didn't eye the street vendor stands on crowded city corners where I used to spend my last dollar on donuts so that I could get one more fix. *Just one more and I'll never do it again.* The restaurants, cafés, grocery stores, and vendors are, indeed, reminders of a past that used to haunt me. Today, though, as I pass them, I can't help but smile, privately acknowledging what I never have to experience again. It's another reminder of how far I've come.

An apathetic taupe doesn't paint my world today. My world is sunny most days. Today, I notice the beautifully tree-lined streets. I sometimes stop to take a picture. When I walk past people, I don't envy those who appear clear-minded and carefree; my mind is clear and my soul is free. I walk the city today with confidence because I know where I'm going and it has nothing to do with food.

While I don't engage in eating disorder behaviors, some of the obsessions still linger from time to time. I still fantasize about eating perfectly. I still have euphoric recall about my skinnier days, thinking about how, someday, I'm again going to fit back into a size two. *I don't want to get too skinny, just smaller. Can you imagine what might happen then? My life would be happy, joyous, and free, and I'd never have to worry about anything!* Eating disordered thinking at its finest.

Today, I know that those fantasies are merely distractions from whatever it is I don't want to face in that particular moment. The difference is that while I might entertain myself with and hear these thoughts, I don't listen to them. I choose, instead, to remind myself that I was never happy when I got to my goal weight or clothing size; with eating disorders, the goal is never attained. I remind myself that I, in fact, felt fatter at a size zero than I have at a size six or size eight. And then I choose to think differently. It's another realization of the insidious nature of the eating disorder.

While my eating disorder has been a daunting obstacle, feeling so unrelenting at times that it was almost too much to bear, I can't say I regret having had it. As much turmoil as the bulimia caused, I can say, in earnest, that I'm grateful for my eating disorder. The eating disorder symbolized my voice, a voice I might not have otherwise found. Not only has it allowed me to think in colors other than black

and white, it has set the stage for me to help others who are struggling, which has become one of my greatest passions.

Looking back at when all of this began—at five years old,
obsessing over homemade butter—and retracing all of my steps as
to how I got here, back in New York City, I realize that I'm exactly
where I'm supposed to be. God (Higher Power, whatever you want to
label it) has had a plan for me all along; I just had to learn to let go
and trust it.

I've made my choices from the buffet of life presented to me—
filling my plate with some items that have tasted amazingly rich,
while other choices have left a bad flavor in my mouth or caused
indigestion. Regardless, the steps of my journey have led me to feel
more grateful than I can ever really convey. Today, my life is full of
experiences, not food. And I have to say . . . it tastes good.

Acknowledgments

T hanks to Janis Donnaud for taking a chance on an unknown writer, like me, and having faith in my story. Without your patience and ability to see my potential, none of this would have been possible. Thanks to my agent, Catherine Crawford, at Janis Donnaud & Associates, for your remarkable and unflinching support throughout the process of publishing my first book.

Thanks to everyone at Seal Press, especially Brooke Warner, for believing in my book. Thanks to Jueli Garfinkle, my amazing editor. I don't know where my book would be without you. Thanks to Robin Bellinger and Willy Mathes for their editorial help when my book was in its infancy. Thanks to all of my friends who read the various iterations of my book and gave me hope to persevere.

Thanks to all of the staff and wonderful people at Sierra Tucson and Turning Point of Tampa. A life-changing experience, indeed. Special thanks to my great friend and Sierra Tucson Alumni Coordinator Tim McLeod, who has been a wonderful presence in my life and helps me stay connected with Sierra Tucson. Special thanks to Vicki Wotring, Mike Hicks, and Lori Herold for your ongoing support.

Thanks to Kay Freyer-Rose, Karen Galey, Dr. Michael Sunich, and Heather Karaman for being such a special part of my journey. I would not be where I am today without each of you. Michael and Heather, I'm so grateful that you both are still part of my life. Thanks to all of the teachers in my life, especially Dawn Dinneen, Linda Muir, Sharon Romine, Nicole Romine, and Loren Whittaker. Special thanks to Dr. Judith Chambers and Dr. Gregory Comnes for allowing me to become a part of your lives outside of the classroom, believing in my ability to write, and being an integral part of my healing process.

I'd like to thank all of my family for their support throughout the years. A very special thanks to my mom, dad, Kent, and Lavada for supporting me throughout my life in so many ways and allowing me to air our past experiences and family struggles in public. Thank you for putting up with me during the turbulent years and not giving up on me. You all have given me so much, to which I will always feel grateful. Thank you for Jan and Lou Merriweather for welcoming me into your family. I'm so glad to be a part of the family.

Finally, a very special thanks to my wife, Lisa, who has unconditionally supported me for the last eight years and who makes me laugh, fills my spirit, and continues to remind me that I'm okay just as I am. Thank you for your patience and sticking by me. I love you.

© KEN BURNS

About the Author

n recovery from bulimia for the last ten years, Greta Gleissner uses
her experience as a platform to help others. Greta has been a con-
tributor for *The Disordered Times*, and *The F-Word*. She is a blogger
for Gurze Books (www.eatingdisordersblogs.com) and also publishes
a personal blog, *Life with Cake,* about eating-disorder recovery. Greta
has a master's in social work from New York University, dedicates her
life to the eating disorder field, and is a member of NEDA, NASW,
and IAEDP. Greta is the Day Treatment Team Leader at The Ren-
frew Center of Old Greenwich, Connecticut, and has a practice in
New York City.

Selected Titles from Seal Press

For more than thirty years, Seal Press has published groundbreaking books. By women. For women.

Purge: Rehab Diaries, by Nicole Johns. $16.95, 978-1-58005-274-0. An honest, detailed account of Nicole Johns' experience in an eating-disorder treatment facility, avoiding the happily-ever-after while offering hope to the millions struggling with eating disorders.

What You Really Really Want: The Smart Girl's Shame-Free Guide to Sex and Safety, by Jaclyn Friedman. $17.00, 978-1-58005-344-0. An educational and interactive guide that gives young women the tools they need to decipher the modern world's confusing, hypersexualized landscape and define their own sexual identity.

Riding Fury Home: A Memoir, by Chana Wilson. $25.00, 978-1-58005-432-4. A shattering, exquisitely written account of one family's struggle against homophobia and mental illness in a changing world—and a powerful story of healing, forgiveness, and redemption.

Addicted Like Me: A Mother-Daughter Story of Substance Abuse and Recovery, by Karen Franklin and Lauren King. $16.95, 978-1-58005-286-3. A mother and daughter share their candid struggles with addiction—thirty years apart—giving readers insight into how to break the cycle.

A Thousand Sisters: My Journey into the Worst Place on Earth to Be a Woman, by Lisa Shannon, foreword by Zainab Salbi. $16.95, 978-1-58005-359-4. Through her inspiring story of turning what started as a solo 30-mile run to raise money for Congolese women into a national organization, Run for Congo Women, Lisa Shannon sounds a deeply moving call to action for each person to find in them the thing that brings meaning to a wounded world.

Girl in Need of a Tourniquet: A Borderline Personality Memoir, by Merri Lisa Johnson. $16.95, 978-1-58005-305-1. This riveting and dramatic personal account gives us a glimpse of what it means to be a borderline personality in a relationship.

Find Seal Press Online

www.SealPress.com
www.Facebook.com/SealPress
Twitter: @SealPress

WHY
HUMANS
LIKE TO
CRY

Tragedy, Evolution,
and the Brain

MICHAEL TRIMBLE

OXFORD
UNIVERSITY PRESS

OXFORD
UNIVERSITY PRESS

Great Clarendon Street, Oxford, OX2 6DP,
United Kingdom

Oxford University Press is a department of the University of Oxford.
It furthers the University's objective of excellence in research, scholarship,
and education by publishing worldwide. Oxford is a registered trade mark of
Oxford University Press in the UK and in certain other countries

First Edition published in 2012
First published in paperback 2014
Impression: 1

Published in the United States of America by Oxford University Press
198 Madison Avenue, New York, NY 10016, United States of America

British Library Cataloguing in Publication Data
Data available

ISBN 978-0-19-969318-4 (hbk)
      978-0-19-871349-4 (pbk)

Printed in Great Britain by
Clays Ltd., St Ives plc

To Peter, who shares a love of music, and with whom
I have spent many hours listening to, discussing, and
enjoying opera, the quintessential artistic expression of
what it means to be human.

Acknowledgements

I wish to thank several people who have contributed their time in helping me with this book, either by reading through sections or the whole text, or preparing and organizing earlier and the final drafts, in particular Jackie Ashmenall, Antonio Damasio, Dale Hesdorffer, Pamela Jencks, Ian Jenkins, Annabel Obholzer, Stephen Porges, Simone Shamay-Tsoory, Anthony Smoker, Tony Verity, and Dame Jenifer.

I am grateful to the late Lennart Heimer for his contributions to my explorations of neuroanatomy and for gifting me many of his neuroanatomical images, and to Nancy Heim who created several of the other illustrations.

Finally I am indebted to Latha Menon, Emma Marchant, Jenny Lunsford, Nicola Sangster, and all those at Oxford University Press who have helped with the preparation of this book.

CONTENTS

LIST OF FIGURES

1

~

INTRODUCTION

Incipit tragoedia

(Tragedy begins)

Nietzsche, *The Gay Science*[1]

In the summer of 2008, Gana the gorilla gave birth to a male baby in Münster zoo, which three months later died of unknown causes. Images of Gana holding on to the dead infant for several days were widely reported in the newspapers and on the Internet. Someone from the zoo said that such behaviour was not uncommon in gorillas, and it was interpreted by another as mourning. This may well have been correct, but what was more interesting was that yet another spokesman implied that the whole of Germany was mourning for her. It was reported in the newspapers that many visitors to the zoo, who came to see Gana and the dead child, were moved to tears—but Gana shed none.

It is obvious that many species have the ability to lachrymate, to shed tears, and this has direct biological relevance, in part related to maintaining a healthy eye. However, the evidence that animals other than *Homo sapiens* shed emotional tears is lacking. Marc Bekoff, in his book *The Emotional Life of Animals* reviews much literature and many anecdotes on animal emotions, but neither 'tears' nor 'crying' appears in the index, and the only tears commented on are his own when looking into the eyes of a cat, or those of others who had been moved at seeing animals suffer.[2]

There are always people who will testify that their dogs cry, and all animal watchers will have seen distressed animals, whose behaviour may be anthropomorphically interpreted as akin to human grieving. Charles Darwin (1809–1882) noted in his *Expressions of Emotions in Man and Animals* that the Indian elephant was known to weep sometimes, and he had heard reports of weeping *Macacus* monkeys, but he had no personal direct observations. When he went to the Zoological Gardens and enquired about the behaviour of two monkeys that were the subjects of the anecdotes, observations of weeping were not confirmed.[3]

The animal most associated with weeping is the elephant, but the conclusions of the book entitled *When Elephants Weep*, another sweep through animal emotions, was that 'in the years since Darwin's observations, the balance of evidence has been the same: most elephant watchers have never seen them weep, or have rarely seen them weep when injured'.[4]

Our closest animal relatives are the apes, but Gana herself did not cry with tears, and there is no good evidence that gorillas or chimpanzees shed tears for emotional reasons. Dian Fossey,[5] who observed gorilla behaviour over many years, reported seeing crying which she thought was associated with grief on only one occasion. Jane Goodall, who documented much chimpanzee behaviour, including their use of tools, their close contact (hugging and kissing), their intimate mother–infant bonding, their warlike brutality even to members of their own species, and their close emotional bonding which can persist for several decades, has not observed chimpanzees crying emotional tears.[6]

There are many reports of animals shedding tears in painful settings (with physical abuse or from natural causes), but this book is not concerned about tears in such settings. It concerns tears that are shed in response to emotion, not physical pain, and is therefore focused on humans. It also does not enter into a discussion of crying associated with psychiatric disorders, especially depression. Certainly crying is but one sign of a depressive disorder, even though in severe depression patients often report that, although very sad, they cannot shed tears. While this is worthy of discussion in its own right, what is of interest here is crying for emotional reasons, in everyday life, especially in response to art. While mood disorders are associated with at least some forms of creativity, these matters go beyond the brief of this book.

In an earlier book, *The Soul in the Brain: The Cerebral Basis of Language, Art, and Belief,*[7] I have argued that there are neurobiological associations that link poetry, music, religious feelings, and affective, especially bipolar, disorders, and gave pride of cerebral space to the non-dominant right hemisphere of the brain, which I concluded is dominant for many human social and cultural behaviours. That book insisted on an evolutionary approach to an understanding of the life and soul of *Homo-sapiens*, and this theme is part of the undercurrent of this present text. Creativity is linked with basic neurobiological forces, the latter sculpting the mind and artefacts of our primate ancestors, culminating in the great artistic achievements of the past few millennia. Understanding the functions of the right hemisphere, until recently relatively neglected by neuroscience, I suggested, at the least gave us an insight into our love of music and poetry, and our religious feelings, indeed into love itself. I also broached the topic of the art form Tragedy—tragic drama releasing in spectators a special emotion, a theme revisited in relation to crying.

In this book, following common practice, I shall use tragedy with a lower case *t* to represent individual experience, while Tragedy with a capital *T* is used for the art form. The traditional idea is that Tragedy developed in classical Greece in the 5th century BC, and refers to the plays of the tragedians, which were seen in theatres, as part of communal religious festivities. Tragedy in this sense is a staged event, with serious intent, within which suffering occurs. Over time, various authors from Aristotle onwards attempted to describe and prescribe Tragedy in ways which simply confined it. These academic arguments go well beyond the scope of this book, but are touched on where necessary in order to understand our emotional responses to viewing Tragedy on the stage, in the cinema, or on television. Crying in such settings brings pleasure, and seems to have been doing so for well over 2,000 years. This strange phenomenon requires an explanation. It has received much comment from philosophy and social science, but in this book I explore the underlying evolutionary neurobiology.

Events leading to great suffering, often combined with a sense of injustice, such as the sudden death of a young person, produce emotional turmoil and are described as tragic. Such tragedy affects many individual human beings. The link between tragedy and Tragedy has been well discussed, but mainly from the point of view of literary criticism, rather than from the point of view of the emotional responses to tragedy and Tragedy. This delicate ground I explore in the book, as well as trying to discover when, from an evolutionary perspective, our ancestors first became aware of the tragedy of life, and when the responses to that involved crying.

I also seek to understand any direct links between tragedy and Tragedy. The polymath Arthur Koestler put it this way:

> every individual lives in a state of 'split consciousness': There is a tragic plane and a trivial plane, which contain two mutually incompatible kinds of experienced knowledge. Most of our lives are lived on the everyday, trivial plane, but in moments of elation or danger, we find ourselves transferred to the tragic plane. The latter, with its cosmic perspective, drowns out for a while the shallow frivolities of life, but they always return, and sometimes the two perspectives coexist.[8]

The book seeks to understand how our emotional responses and crying for emotional reasons have evolved over several millions of years, and also to explore why it is that emotional crying is a uniquely human attribute. This entails an exploration of the neuroanatomy of the human brain, seeking differences from those of other primates, especially with regards to the representation of emotion, and the circuitry related to the release of tears. At first sight, this may seem a daunting task; neuroanatomy is complex, even to those with medical training. It is a challenge to enlighten non-neurological readers with the intricacies of the brain, without being either so simple as to irritate those with some previous knowledge, or so elaborate as to lose those who are interested but to whom the various names involved seem like a foreign language. In order to help with this, I have prepared an appendix (1), which is a brief overview of the anatomy involved in Chapters 3 and 4. For the uncertain, or faint-hearted, it is recommended that this is read after Chapter 2, and before tackling Chapter 3. As a further aid, there is a glossary of terms to be found in Appendix 2.

Before we look into neuroanatomy, though, it seems appropriate to say a few words about Friedrich Nietzsche's seminal text *The Birth of Tragedy*, published in 1872, in which he explores the connections between the Tragedy of the Greek dramatists and the cultural and religious contexts in which they were writing. We will be touching on Nietzsche's work at various points in this exploration of crying and Tragedy.

NIETZSCHE, SCIENCE, AND *THE BIRTH* OF *TRAGEDY*

Three aspects of Nietzsche's theories are pertinent to this book. First, note the full title of the book, *The Birth of Tragedy: Out of the Spirit of Music*: music is central both to his thesis and to the solutions of the questions posed above. Secondly, note his introduction of the interplay between the two iconic Greek gods Dionysus and Apollo. With the added twist of neuroscience, these much discussed images may be seen in something of a new light, as metaphors for psychological processes based in neuroanatomical and evolutionary principles. Thirdly, and much less commented on, is Nietzsche's use of physiological and psychological ideas that were current at the time.

Nietzsche (1844–1900) grew up surrounded by music. In his lifetime he knew many skilled professional musicians, and he was also endowed with musical gifts. At the time the book was published, Nietzsche was professor of philology at Basle University and had become well acquainted with the composer Richard Wagner (1813–1883). Nietzsche and Wagner were both at one point influenced by the philosophy of Arthur Schopenhauer (1788–1860), who viewed music as very special among the arts, an expression of what he referred to (rather unfortunately, given the ambiguity of the term) as the Will in its direct manifestation.[9] Music, because of its non-representational nature (i.e., it does not deal with things in the real world) and being independent of the world of phenomena, was able to reveal truths about the essence or intrinsic nature of things, and thence life.[10] The world we see (in our perceptual field), is only a representation of what is actually out there; it is only appearances—in neuroscience terms, visual information filtered through the filigree of our nerves and brain. Other forms of art, such as painting or sculpture, re-represent the visual experience, offering the personal mould of the artist to the viewer or listener. But these cannot, in Schopenhauer's scheme of things, get underneath the illusion to the thing-in-itself, to touch the Will. This is arguably the pulsating energy that invests all creation, what Schopenhauer himself called 'the essence of everything in nature...eternal becoming, endless flux'.[11] Following on from Schopenhauer, music was thus seen by Nietzsche as Dionysian: it escaped the phenomenon of individuation.

The Birth of Tragedy begins with the two gods Apollo and Dionysus and their fundamental importance to the cultural development of Western art. Apollo and Dionysus together give birth to Greek or Attic Tragedy, which emerged out of the spirit of music: 'art derives its continuous development from the duality of the *Apolline* and *Dionysian*'.[12] Apollo represents the rational, the world of reason, the slayer of chaos. He is the deity of light and represents *individuation*, separating us from each other and dividing up things in our phenomenal world.[13] Apollo, epitomized by Doric art, especially sculptures, and reflected by the epic in poetry, is the god who builds boundaries. Dionysus, in contrast, claims melody, lyric poetry, and the dance. The god of wine, of ecstasy and rapture, Dionysus encourages the rupture of boundaries.

8

Nietzsche's theory of Tragedy contains a fusion of Apollonian beauty with Dionysian energy and sadness, with Dionysian wisdom penetrating through Apollonian artefacts. He laments the weakening of Dionysian creative energy in art, as over time Apollonian form and order have constricted its breath and breadth. With this burden, Tragedy, that is, true Tragedy for Nietzsche, disappeared from the stage. This he blamed on the rise of a Socratic culture, seeking truth through knowledge, driven by dialectics and what he called 'scientific optimism'. The Greek philosophers Socrates and Plato in their writings both repudiated Tragedy, and for Nietzsche the Socratic dialectic was simply too optimistic in its view that rational minds could solve the problems of existence: it was inherently anti-tragic.[14] Dionysus was laid to rest, Apollo became ascendant, and the systematizers and classifiers took over. 'While in all productive people instinct is the power of creativity and affirmation, in Socrates instinct becomes the critic, consciousness the creator—a monstrosity *per defectum*!'[15]

Nietszche refers to the descendants of Socrates as 'at bottom a librarian and a corrector of proofs, wretchedly blinded by the dust of his tomes and by printing errors'.[16] Rationality rather than instinct came to dominate Western culture but for him, decadence had set in with Socrates and his ilk, rather than, as is usually portrayed, with Dionysus and the forces of chaos. Socratic consciousness and Platonic dogmatism supplanted Homeric action; the hero died not of Tragedy, but of a failure of logic. Socrates, who was put to death for corrupting youth, was seen by Nietzsche as life-denying, since his way of

seeing things led not only to the fostering of a slave mentality—individuals being subservient to a set of external rules—
but what was worse, rules for which there could be no
validation.

The two central mythological gods of Nietzsche's *Birth of
Tragedy*, Dionysus and Apollo, are not to be taken literally, and
are used by him as emblems for his theories. Nietzsche later
rejected some of his earlier ideas, especially his views on the
kind of pessimism advocated by Schopenhauer, and modified
his adoration for Wagner and his music. In 1886 he produced
a second edition of the book, but now renamed *The Birth of
Tragedy; or Hellenism and Pessimism*. He referred to the original
edition of the book as 'questionable...impossible...a youthful work full of youthful courage'.[17] He accepted some of the
criticisms that had been levelled at it after its publication,
especially its lack of style and of historical precision, even the
tenuous classical scholarship. But he remained faithful to 'the
tremendous phenomenon of the Dionysiac...an "unknown
god", disguised beneath the scholar's hood...a spirit with
strange, as yet nameless needs, a memory bursting with questions, experiences, mysteries'.[18]

Nietzsche apologized for trying to tackle such a difficult
psychological question as the origin of Greek Tragedy, and
enquired about the possible *physiological* meaning of Dionysiac madness.[19] But he now identified Dionysus with the Anti-
christ, and the polarity with Apollo is no longer apparent.
Christianity is charged with denying life, hating the world,
fearing beauty, and condemning the emotions. The place of

music also appears to have been diminished in his ideas, if the new title is anything to go by, revealing his disappointment with Wagnerian romanticism.

Nietzsche discussed the book again in *Ecce Homo*, written in 1888, an autobiographical text discussing some of his earlier works. Schopenhauer's pessimism was totally rejected—if anything, Nietzsche said of his earlier work on Tragedy that it smelt offensively Hegelian, but Dionysus was still there.[20] He reaffirmed the importance of the Dionysian phenomenon, which offered a psychology, but one symbolizing affirmation of the totality of life, which remained in his thought as a basis of Hellenic art and tragedy. Within the development of his ideas he set forth a philosophy of *becoming*, in contrast to being, since change, transience, and destruction were elemental to the Dionysian.

The archetypes created by Nietzsche were borrowed from the Greeks. The book came in for considerable academic criticism, since Greek gods are much more complex and have a compass of many attributes. For example Apollo, rather than Dionysus, was the god of music, albeit of the lyre, not the aulos or reed pipe, and of poetry, albeit of the epic not the lyric. In their comprehensive analysis of Nietzsche and Tragedy, Michael Silk and Joseph Stern note that neither image has ancient authority, although Nietzsche's Dionysus, while borrowing much from the Orphic religion, was closer to the 'real' Dionysus of the Greeks. What is important in the context of this book, however, is not so much their apparent opposition but their synthesis, and their *psychological* as opposed to any classical or theological meaning. In addition, there is Nietzsche's use of the word 'physiological'. Form under the tension of force, the artistic articulations of Apollo shaping Dionysian energy, as the tragic hero as individual is over-whelmed by forces and circumstances, natural and social. Tragedy is revealed as a narrative of time, of life and of becoming, as 'artistic powers which spring from nature itself', as Nietzsche put it.[21]

Nietzsche was interested in science, and it is known that his thinking progressed through various stages, being rather more influenced by contemporary scientific writings at one time than at others. Although he was not a Darwinist, and evolution is not mentioned in *The Birth of Tragedy*, he lived through an age when *Origin of the Species* was published (1859) and discussed. There was much contemporary scientific discourse on the forces of nature and the relationship between force and matter. Further, he was briefly at the battlefront in the Franco-Prussian war, experiencing death and destruction, power and energy, at first hand. Dionysus was representative for Nietzsche of creative and destructive forces, necessary to each other, and of power, a theme which became central to his evolving philosophies.

These are some of the words he uses throughout the text of his book: 'drives', 'wills', 'energies', 'impulses'. The German word *Trieb* is frequently employed, referring to desire and instinct.[22] Apollo and Dionysus are personified as driving forces (*Kunsttrieb*), physiological phenomena. John Sallis suggests that such ideas hint at states of nature that have

INTRODUCTION

anticipated art: 'It is as though nature itself already contained the transition from nature to art, even if holding it in a certain reserve; or, rather, it is as though nature and art crossed in a region that would be neither simply nature nor yet art, a kind of proto-art or proto-artistic nature.'[23] It is grounded in a pre-conceptual level of response. Aesthetics, wrote Nietzsche, is 'nothing but a kind of applied physiology'.[24]

In his later philosophical development, Nietzsche acknow-ledged our animal inheritance, and the needs of our cognitions to serve survival. Along with the physiological, 'intoxication' (*Rausch*) is another theme. Nietzsche even wrote about the brain. He referred to the creative act as associated with 'the cerebral system bursting with sexual energy'.[25] More interest-ingly, at another time, he referred to 'two chambers of the brain, as it were, one to experience science and the other non-science: lying juxtaposed, without confusion, divisible, able to be sealed off; this is necessary to preserve health. The source of power is located in the one region: the regulator in the other. Illusion, partialities, and passions must provide the heat.'[26]

Nietzsche's approach to evolution has been discussed in some detail by John Richardson in *Nietzsche's New Darwinism*, and in Gregory Moore's *Nietzsche, Biology and Metaphor*.[27] With-out expanding on their theses, which in part revolve around the development of Nietzsche's concept of Will to Power, we can note the following points. Nietzsche, while being hostile to Darwin's ideas, even though he never read his works first-hand, has been accused of not understanding them properly. In his own biology, Nietzsche places the human, from an evolutionary perspective, back among the animals, whose evolved forms are underpinned by the notion of drives. Nietzsche criticized Darwin and the idea of natural selection on the grounds that the latter placed too much emphasis on the extrinsic as opposed to the intrinsic dynamics of change. Natural selection alone was in his view too passive an agent to explain evolutionary development.

Drives embedded in nature were not generally considered conscious, and yet 'they weave the web of our character and our destiny'.[28] They are part of our physiology. But our cognitive structures are also embodied, as we are embedded in our social world. Will (as a psychological drive underpinned by physiological forces) is seen as a product or as part of the evolutionary process. Consciousness itself also arises from physical, biological processes, as does the development of our valuations (including moral ones). In Nietzsche's philosophy, mankind represents an animal whose nature has not yet been fixed.

Nietzsche's views implied a new way of viewing the relationship between perception and emotion, and to a psychology in which the emotions dominate, based not on the mind–body dualism of Descartes (1596–1650), but on Apollo and Dionysus.[29] These drives are organic, and in Richardson's view are closely linked with *Rausch*: they 'quicken the organism…aesthetic experience as characterised by this "visceral" excitement or heightening'.[30] They are creative as well as pleasing, a strength that leads to enhanced sensitivity to sensory stimuli, and takes us eventually to dance and the theatre.

Nietzsche was aware of the large subterranean edifice of the human unconscious, and his writings anticipated those of Sigmund Freud (1856–1939). Many of Freud's ideas spring out of the pages of Nietzsche, but Freud declined to acknowledge this, remarking that although he had Nietzsche's books in his library, he had not read them for fear they would interfere with his own ideas![31] Consciousness for Nietzsche was but 'a fanciful commentary on an unknown, perhaps unknowable, but felt text'.[32] Consciousness, he opined, belongs to the herd nature of man, it is an animal consciousness which 'developed only under the pressure of the need for communication...conscious thinking takes the form of words'.[33]

Nietzsche's polarization of the two gods was a theme taken up by others, notably Thomas Mann (for example, in *Death in Venice* and *Dr Faustus*), Hermann Hesse (in *Narcissus and Goldmund*) and Henrik Ibsen (in *Emperor and Gallilean* and *Hedda Gabler*). Camille Paglia, giving little credit to Nietzsche, contrasts Apollo, the lawgiver and representative of sculptural integrity, with Dionysus, god of fluids, of *sparagmus*, and of pleasure-pain.[34] Her thesis is that the Apollonian and the Dionysian are two great principles governing the sexual personae in art and also in life:

> Apollo, is the hard cold separatism of Western personality and categorical thought. Dionysus, is energy, ecstasy, hysteria, promiscuity, emotionalism—heedless indiscriminateness of idea or practice...Complete harmony is impossible. Our brains are split, and brain is split from body. The quarrel

between Apollo and Dionysus is the quarrel between the
higher cortex and the older limbic and reptilian brains.[35]

In seeking the roots of emotional crying, we will look at
brains, bearing in mind the evolutionary forces which have
shaped our human neurobiology and social structures. While
Nietzsche's views on evolution, forces, and the Will have been
subject to much criticism, their imaginative provocative ideas
have resonance today in at least some philosophical circles.
However, they also have relevance for neuroscience, especially among thinkers and researchers exploring cognition,
emotion, and consciousness.

I have reached the conclusions in this book over many
hours of watching and discussing Greek Tragedy, listening to
music (especially opera), and having studied the neuro-
sciences all my professional life. One conclusion is that music,
above all the arts, is simply special, with effects on us above
and beyond the other arts. Some may disagree, but it appears
to be the art form most likely to make us cry. I submit that
emotional crying is a unique attribute of *Homo sapiens*. Which
one of us could call ourselves human if we did not sometimes
cry; if we did not respond to beauty, to suffering, or to the
death of another with tears?

CRYING

Sunt lacrimae rerum et mentem mortalia tangent

(The way things are calls for our tears and
mortality touches our hearts)

Virgil, *Aeneid*[1]

A fter the fall of Troy and the end of the Trojan War,
Aeneas arrives at Carthage on his way to Italy to found
a new city. At Juno's temple he sees painted images of the
fallen heroes of the war, and he weeps.

Crying as an emotional response, especially to sadness and
bereavement, is portrayed in the earliest of Western literature,
and there are many descriptions of it in Homer's *Iliad* and
Odyssey. Great heroes weep. Odysseus sheds unconstrained
tears as the sacred singer Demodocus sings songs about the
Trojan wars; Achilles cries at the death of his friend Patroclus;

and even the great Agamemnon breaks down in tears when embracing Odysseus, who visits him after his death in the underworld. As with Aeneas, these heroes weep for the tragedy of the loss of friends and companions during war. Odysseus blames the gods for the whole catastrophe, for weaving tragedy into men's lives. Such suffering was commemorated in songs that formed the essence of Dionysian theatre: the communal song, the hymn, and the core of Tragedy.

CRYING AS AN EMOTIONAL RESPONSE

There are many observations of animals that cry out, vocalize, and express distress, which seem to reflect the human equivalent of pain or bereavement, but crying seen as tearful sad sobbing seems to be a distinctly human attribute. This is not to say that animals do not feel what we may call sadness or sorrow, or even mourn in their own way. Emotional contagion and empathic responses have been well defined in several species, but tearing in such situations seems not to be within their experience.[2] A singular problem with the observations is the anecdotal and anthropomorphized interpretations of the behaviours which are viewed by other scientists as insufficient evidence of the true nature and spectrum of animal emotions. This is compounded by the difficulties of defining human emotions adequately. However, the idea that animals do not have emotional experiences, or that the ones they have are in some sense not equivalent to human feelings, would seem to ignore the neuroanatomical fact that brain structures subserving emotion are found way down the evolutionary scale, and become well developed in mammals, as described in the next chapter.

Emotional weeping is not only uniquely human, but universal. Tom Lutz in his book *Crying* surveys the cultural aspects of tears, and comments on their centrality in works of art across the ages. He notes that the first recorded instance of tears is found in Canaanite clay tablets (14th century BC), in which there is an account of weeping at the news of the death of the ancient Semitic god Ba'al by his sister and lover Anat. In Egyptian mythology, Isis weeps for the dead Osiris, and in the early Mesopotamian epic *Gilgamesh*, considered one of the first works of literary fiction, the hero-king Gilgamesh mourns for his companion Enkidu with tears that last seven nights.[3]

The conflation of crying with religious themes is a part of this tradition. There are tearful lamentations in hymns ('out of the deep I cry to thee: oh Lord God hear my crying'[4]) and psalms (passing through the valley of tears[5]): there are the bloody tears of the saints, and the religious statues dedicated to them. St Francis of Assisi went blind, probably as the result of a trachomatous infection of the eye, but this was, and is still by some, attributed to excessive crying—he is believed to have cried his eyes out. He reflected that he preferred purifying his spiritual vision with floods of tears to going blind, but seems to have succeeded at achieving both. Weeping at the Wailing Wall in Jerusalem seems obligatory—such holy tears are seen to be stigmata of a true love of God.

WHERE AND WHEN

William Frey carried out an extensive study in the 1970s on the epidemiology of emotional crying, on 331 adults without psychiatric problems, from a variety of social settings, including a sample of twins. He observed that the average frequency of crying was 5.3 times a month for women, compared to 1.4 for men. The scatter was wide, between those who cried daily and those who were tearless. Six percent of the women but 45 percent of the men did not cry during the recorded month. Of considerable interest is that people reported feeling happier after crying. In this study, 85 percent of female and 73 percent of men reported feeling better after shedding tears; some reported that the act relieved tension, others commented on the cleansing action of crying—washing out bad feelings and similar sentiments. The subjective feelings, however, were dependent on the situation. Crying after a domestic dispute or after emotional traumas which threatened life and limb was not associated with such positive feelings, and the episode of crying was inclined to last a much longer period of time.

Sadness was the primary emotion linked with tears in men and women, and crying induced by sadness was of longer duration than tears of joy. Also, the act of crying was reported as often being preceded by the feeling of a lump in the throat.

These results have been in the main replicated by others, although later, more sophisticated studies have emphasized the cultural and contextual variability of crying: settings and companions modify the response, as does the gender of the person crying.[6] Jeffrey Kotter reports that males are less likely to use tears manipulatively, and that they cry in more subtle ways compared to women. By this he means that they shed fewer tears and for a shorter duration, they are not inclined to explain why they cry, and they apologize more for the tears. He goes on to observe that men tend to cry in response to specific situations, and only two of these are equivalent to those in which women cry, namely at the death of a loved one, and in relation to a moving religious experience. Men, he argues, look more towards internal as opposed to external cues, and cry over feelings that relate to their core identity as providers and protectors, as fathers and fighters.[7]

Ad Vingerhoets and his colleagues have explored the question as to when the effect of crying is reported as cathartic.[8] The word 'cathartic' has permeated the literature of Tragedy and crying for many years, and is discussed in more detail later, but it refers to a supposed relief of emotion that occurs in certain settings. The investigators carried out a survey in the 1990s of crying in 2,181 men and 2,915 women from 35 different countries. The study involved the use of assessment scales, such as the Adult Crying Inventory, but also investigated mood and physical changes, gathering data in relation to the most recent crying episode.

Feelings of loss were the most frequently reported antecedent. After crying, most respondents recorded an improvement in their mental and physical state, although this was not the case for some 10 percent, who felt worse mentally, and a greater percentage who were worse physically. There were no significant gender differences for the mental state improvements after crying, although men were less likely to report feeling physically worse.

The presence of others and the social context in which the crying occurred was important, and crying alone or in the presence of one other person led to the greatest 'cathartic' effect. The latter was also associated with the presence of social support, the resolution of any event that was associated with the crying, and an enhanced understanding of the circumstances that had led to the crying in the first place.

Although it is often reported as associated with prior feelings of loss, crying is generally viewed in a positive way, and any negative views, in Frey's survey, were related to an association with 'weakness', especially in men. Odysseus, who cries on hearing the laments of Demodocus, weeps, as Homer put it, like a woman, and hides his head and face with his mantle, ashamed to be seen crying by the assembled Phaeacians.[9]

The one-time myth that grown men don't cry has been broken by the many recorded instances of public tears. Big boys as well as heroes cry. Theodore Roosevelt was seen to cry in public, and Jeffrey Kotter quotes other US presidents who have done the same, including Ronald Reagan, Bill Clinton, and even George Bush, senior.[10] The famous baseball player Babe Ruth cried when he announced that he had cancer, and the boxer Floyd Patterson did so after losing a fight to Muhammad Ali. Male watersheds in the movies include

Marlon Brando in *A Streetcar Named Desire*, Tom Cruise in *Magnolia*, Russell Crowe in *Gladiator*, Leonardo DiCaprio in *Catch Me If You Can*, George Lazenby in *On Her Majesty's Secret Service*, and the great heart-throb James Dean in *Rebel without a Cause*, in his case with no apparent cause—just a troubled soul in an indifferent world. John Wayne is said to have jibed that he would cry for his horse, his dog, or a friend, but never for a woman. Male tears shed in public are perhaps more common nowadays than in previous generations.

Kotter refers to good and bad tears, depending on the circumstances of the crying and how the feeling settles after the crisis. Tears are a universal accompaniment of mourning, observed in all societies where it has been studied, and people cry openly in the theatre and the cinema. Kotter describes the tear ceremonies of the Bosavi people, observed by the anthropologist Edward Schieffelin. They systematically work up to weeping on certain occasions, for example, for the entertainment of important guests. There is dancing and singing, burning of the shoulders of the dancers, and howling and weeping through the night. Afterwards, the contented guests apparently pay compensation to the hosts for having made them cry. Kotter suggests this ceremony is about nostalgic Tragedies, likened to our own attendance at a play or opera. They are 'institutionalised tear ceremonies...that help us to reflect on our feelings about our own existence, through the lives of others. Songs and dances tell stories of lost love, making us cry...nowhere is this more evident than during times of death.'[11] Such descriptions remind us of the way in which

Tragic drama serves to release emotions in us, and the importance of crying in social interactions.

WHY?

Over time there has been much speculation about the purpose served by emotional crying. A problem with these theories is that they remain largely theories, entangled with the
complexities of the overdetermined nature of any individual
act of crying. Ancient ideas that held that tears were one way
to get rid of bad humours continue to surface. For example,
Frey's assessments of the chemical constituents of tears led
him to the view that noxious chemicals, which build up as a
result of stress, are removed from the body in crying, literally
an excretory process: purgation by another means. This has
some associations with the theory of catharsis, a view that is
linked to purification and cleansing.

In a slightly different version of these ideas, it has been suggested that tears drain off excess emotional energy, restoring a
homeostasis. This was a favoured theory of the early Freudian
pre-psychoanalytic theories. In Freud's collaborative work with
Josef Breuer (1842–1925), the word 'catharsis' first appears in
Studies on Hysteria, where it is explained that an injured person's
reaction to a trauma is cathartic only if it is complete, such as
in revengeful action. It is through language, as a substitute for
action, that the effect can be modified when incomplete—these
were ideas that were to lead from the cathartic to the psychoanalytic therapies.[12] Kotter, a therapist himself, refers to crying

as a 'defence against other internal drives'; it is an act of regression and a retreat to the earliest preverbal stage of life.[13]

While there is good evidence that crying makes people feel better, there is little evidence showing any cathartic effect of crying, if by that is meant some sort of peaceful relief from tension or another emotion. No cathartic effect of crying has been observed when people are asked to cry as opposed to suppressing their tears while watching sad events.[14] James Gross and colleagues showed a sad film to 150 women and measured a number of physiological and behavioural indices noting differences between those who cried and those who did not. They were also able to compare measurements in the pre-cry phase with the actual crying spell. Crying, associated with self-reported experiences of sadness and pain was distinguished by increased heart rate, increased skin conductance, decreased breathing rate, and increased somatic activity.

They discussed the implications of their findings for two different theories of crying, the *physiological recovery hypothesis*, which implies a restoring of homeostasis akin to catharsis, and the *physiological arousal hypothesis*, which implies increased emotional activity. Their results favoured the arousal hypothesis, but they cautioned that the main effects of crying may not have been observable over the short time intervals of their study, and that the 'catharsis' may occur later.[15]

Lutz reviewed all the evidence available to him, including psychotherapeutic and psychoanalytic studies, and concluded that there was no hard evidence for a cathartic effect of tears, even for so-called cathartic therapy, in which patients are asked to recall as vividly as possible their traumatic experiences. Randolf Cornelius likewise concluded that, in contrast to a catharsis, 'crying is associated with increases in arousal, tension and negative affect... Crying does also not appear to be necessarily beneficial to one's health, as the cathartic model of crying would predict.'[16]

If crying is not physiologically beneficial, what then is the purpose of emotionally aroused tears? Is it entirely psychological? Recurrent sociological interpretations emphasize the communicative value of crying. Crying, like a shout or a sneeze, attracts the immediate attention of others. Tears provoke an emotional response in the observer which, in the more sceptical views, not only elicits sympathy but acts as a manipulative tool. As Shakespeare put it:

> And if the boy have not a woman's gift
> To rain a shower of commanded tears,
> An onion will do well for such a shift.[17]

Several surveys have confirmed that women cry more than men, but this difference is not observable in infants and it becomes apparent only around puberty. In babies, crying has been referred to as an acoustic umbilical cord: the instant appeal to a mother of a baby's crying is obvious.[18] This has been discussed in the psychoanalytic theories of John Bowlby, crying being a part of attachment behaviour; in a child it is a signal of the departure of his or her parents, especially the mother—essentially separation.[19] Kotter also views crying as a means of bonding between individuals, but of all ages, and suggests it is a powerful way of obtaining help and emotional support.

Around the age of three to four months, the relationship between an infant and the surrounding environment takes on a more organized communicative role, with greater self-regulation, and crying becomes tailored, with more specific interpersonal purposes. There are few studies of the neurological accompaniments of tearing in infants, but around this time the electroencephalogram shows distinct changes, with increased synchronous activity.[20] Brain imaging studies have shown the activation of certain brain areas linked with emotion (to be discussed later) in mothers hearing infant crying compared to simple noise.[21] The meaning of these activations is unclear, but they imply a cerebral interplay between infants and mothers in relation to crying.

Crying in a baby leads the mother to pick it up and offer the breast or some other means of pacification. This theme was put into an evolutionary perspective by Paul MacLean, whose work on the organization of emotion in the brain was stimulated in the 1950s by clinical observations of emotional changes in some people with epilepsy. As a comparative anatomist, MacLean viewed animal behaviours as evolutionary adaptations of the brain. He was one of the pioneers of understanding the circuits in the brain which are involved in emotions and expression, a theme taken up in the next chapter. In his original descriptions of the limbic system—the brain structures linked to emotion—MacLean noted an organization stretching back to our reptilian past. He used the expression 'the triune brain' to explore three different major components of the mammalian brain, which he referred to as the proto-reptilian, the limbic, and the neocortical. In his scheme, the limbic system evolved alongside the developing social complexity of the mammals.

He attributed to the limbic structures (which will be described in greater detail in the next chapter) three key mammalian behaviours: nursing and maternal care; audio-vocal communications, vital for maintaining maternal-offspring contact; and play. Key to MacLean's ideas was that 'the history of the evolution of the limbic system is the history of the evolution of the mammals, while the history of the evolution of the mammals is the history of the evolution of the family'.[22] In other words, the development of the limbic system was an essential prerequisite for the development of certain characteristic mammalian behaviours. A characteristic of a young mammal is its need for the presence of its mother, and everything that involves. A reptile hatching from an egg must not cry out for its mother, or else it will be readily detected by predators and eaten. In contrast, a mammalian infant depends on the separation cry for succour and security. If there is no cry, the infant will not survive. The development of family, sibling, and later peer-relevant behaviours, including emotional and interpersonal bonding, correlates with and is related to the evolutionary development of these limbic structures.

Other ideas about crying fluctuate between the sociological and the biological. Darwin noted that the main expressive movements during crying (and other actions such as laughing or blowing the nose) lead to a rise of pressure in the chest and abdomen, which leads to increased blood pressure in the eyes. In order to prevent damage to the eyes, the muscles around them contract. Darwin considered that this protective contraction 'was a fundamental element in several of our most important expressions'.[23] In infancy, screaming leads to engorgement of the blood vessels in the eyes, and the latter leads to contraction of muscles around the eyes, to protect them from the resulting increased pressure. Tears were a reflex response of the lachrymal glands to these events. The act of contraction of the muscles around the eyes caused alterations in the activity of several facial muscles around the mouth, increasing the expression of the gesture. Darwin observed that young infants before the age of two to four months cry out violently, and have their eyelids firmly closed, but even though their eyes become suffused with tears they do not shed them.[24] Supporting a view that the primary function of tears is to lubricate the eyes and nostrils (to aid smell), and that irritation of the eyes leads to stimulation of the lachrymal glands, Darwin suggested that with evolutionary time, the slight irritation became enough to lead to the free secretion of tears. In the human, by later childhood, these reflexes become habitual, being evoked by lesser circumstances than those that arouse the infant: the habits associated with screaming in children become linked to suffering and the relief of suffering. For Darwin, habitual actions can become hereditary and he reasoned that, in comparison to many other emotional gestures, 'weeping probably came on rather late in the line of our descent; and this conclusion agrees with the fact that our nearest allies, the anthropomorphous apes, do not weep'.[25] He concluded, with regards to the pleasure of crying, that 'by as much as the weeping is more violent or hysterical, by so much will the relief be greater—on the same principle that the writhing of the whole body, the grinding of the teeth, and the uttering of piercing shrieks, all give relief under an agony of pain'.[26]

Ashley Montague observed that in infants who cry without tears, the mucous membranes of the nasopharynx quickly desiccate, an effect which is harmful to the delicate cilia and secretary cells of the nasal passages, which would increase the chance of infection. Since tears run from the eye into the nasolacrymal ducts, he suggested that they serve as an adaptive trait counteracting the damaging effect of tearless infant crying. Crying evolved in humans as opposed to other primates because of the prolonged period of postnatal development, and, in view of the antiviral and antibacterial constituents of tears, the more the youngster cries the healthier he or she is. Taking his theory further, he observed that men have larger nasal passages than women, hence the greater amount of overflow down the female face: males blow their noses while females blub.[27]

From crying with tears comes weeping in sympathy, a social response which, as will be argued later, is conditioned or underpinned by a neurobiological basis linked to empathy. Crying as an emotional response is evoked in many settings.

Some people cry at the slightest emotional wave, while for others a storm is needed before the flood. Some personality styles seem more prone to instant tearing than others; the hysterical can be contrasted with the obsessional. The impressionistic hysteric, responding to the immediacy of every situation, characteristically emotionally labile, and easily perturbed by the slightest emotional breeze, seems the opposite of the highly obsessional, whose intense control over the release of feelings is bounded by emotional and muscular rigidity.[28]

In one study of personality and crying, the circumstances of crying in the previous year were rated, and personality questionnaires filled out by 70 male and 70 female volunteers. The death of a friend and breaking up rated highest in terms of occasions. Women cried more frequently and intensely than men, and in both sexes crying positively correlated with personality variables related to empathy. In men, but not women, neuroticism was positively correlated and masculinity negatively correlated with crying.[29] In an unpublished study quoted by Vingerhoets and Cornelius, in which empathy was measured in nearly 500 adolescents, proneness to crying and empathy were strongly correlated in both boys and girls, while in another study empathy was associated with crying but only in females.[30] Reviewing the available literature on personality and crying, Vingerhoets and colleagues concluded that positive associations could be claimed between crying proneness and neuroticism, extroversion, and empathy.[31] As will become clear later, the link between crying and empathy, confirmed by other investigators, is important in understanding human emotional responses, and is echoed by an underlying neuroanatomy.

Koestler referred to the 'logic of the moist eye' in *The Act of Creation*. He noted different situations in which weeping occurred. Rapture was self-transcending, which led to quiescence, tranquillity, and catharsis. No specific voluntary action could consummate the moment; as he pointed out, you cannot take a stunning visual panorama home with you—as every photographer knows. He equated this response with altered activity of the autonomic nervous system (an anatomical concept to be discussed in Chapter 3). When weeping in sympathy with another, or on viewing a tragic film, he suggested that two psychological processes occur: identification, which he equated with introjection or empathy, and vicarious emotions.[32]

Thus, catharsis, in a physiological sense, has been difficult to substantiate, but the results are by no means conclusive. More work on the delayed responses to crying, in which the aftermath of the tearing is evaluated in comparison with the states before, needs to be carried out, and more sophisticated studies using, for example, the newer methods of brain imaging could be rewarding. However, the surveys reveal that with some occasions of crying people experience what they call a cathartic experience, and that the feeling is usually positive.

Some people feel so much grief that they simply cannot cry. In the play *Titus Andronicus*, Shakespeare explores a grief that exceeds tears. In *The Tin Drum*, Gunter Grass describes the Onion Cellar, a bar in postwar Germany where the guests are given only onions and knives. The seven skins of the onions are peeled away, the onions are then chopped, and people no longer see anything because their eyes overflow with tears. People go to the cellar just to share painful memories and cry. The protagonist reflects how 'the juice' brought forth what the world and the world's suffering could not: 'a round human tear...the rains came, the dew fell...the tragedy of human existence was spread fully before [them]'.[33]

AESTHETICS, CRYING, AND TEARS OF JOY

Some of the more psychologically based theories of the effects of crying have emphasized cognitive aspects, in other words the evaluation of context within which crying occurs and its interpretation for the individual. Kotter refers to the language of tears, with its own syntax and grammar. This language transcends words, and has its own rules and a unique vocabulary. Tears authenticate meaning: they reflect honesty; they hide so much that cannot be said in words and so little. So much in the sense that they sometimes reflect times, places, and people embedded in the autobiographical memory that are perhaps unavailable to the consciousness of the person who is crying, as well as of those who witness the crying. So little in the sense of the powerful communicative value and meaning of crying. Tears have a symbolism, for the one crying and for those who observe the tears; they are woven from fragile, mutable memories with more than a hint of mortality.

Antiquity itself has such resonances. It can be moving to touch the old and bygone, to be touched by the past in the present. Proust, in *The Guermantes Way*, describes an old archaeologist in Berlin of whom it is said that

> if you set him in front of a genuine piece of Assyrian antiquity, this old archaeologist weeps. But if it is a modern fake, if it is something that is not really old, he fails to weep. And so, when they want to know whether an archaeological piece is really old, they take it to the old archaeologist. If he weeps, they buy the piece for the museum. If there are no tears, they send it back to the dealer and prosecute him for fraud.[34]

Here is William James (1842–1910) on aesthetics: 'In listening to poetry, drama or heroic narrative we are often surprised at the cutaneous shiver which like a sudden wave flows over us, and at the heart swelling and the lacrymal effusion that unexpectedly catches us at intervals. In listening to music the same is even more strikingly true.'[35] Shelley proclaimed the power of music in this way:

> I pant for the music which is divine,
>   My heart in its thirst is a dying flower;
> Pour forth the sound like enchanted wine,
>   Loosen the notes in a silver shower;
> Like a herbless plain, for the gentle rain,
>   I gasp, I faint, till they wake again.[36]

Kotter, James, and Shelley, from their different perspectives, remind us that combinations of sights and sounds are sensational experiences, which in harmonious combinations may

arouse secondary pleasures, enhanced by memory and association. They imply a bodily sounding-board, which vibrates
within us, and which can shake the soul. The importance of
music in these aesthetic responses will be returned to in a later
chapter, but has been discussed by Leonard Meyer in his influ-
ential book *Emotion and Meaning in Music*.[37] He notes that, in the
main, musical theorists have concerned themselves with the
grammar and syntax of music rather than with the affective
experiences that arise in response to music. Music, if it does
nothing else, arouses feelings and associated physiological
responses, and these can be measured. Meyer points out that
the response to a piece of music need not be conscious, but
those trained in music tend, because of the critical attitudes
which they have developed in connection with their own artistic efforts, to become self-conscious of their aesthetic experiences, to objectify their meaning, and to consider them objects
of conscious cognition.[38] Although Meyer was uncomfortable
with the idea that music was a universal language, pointing
out the transcultural and transhistorical variants of style, he
nevertheless reflected that 'these languages have important
characteristics in common...Most important is the syntacti-
cal nature of different styles...the organization of sound
terms into a system of probability relationships, the limitations imposed upon the combining of sounds etc....In this
respect, musical languages are like spoken or written languages which exhibit common structural principles.'[39]

Meyer also discussed the relevance of memories for the affective musical experience, and how these are evoked by either

conscious connotation or unconscious image processes. It is
these imaginings that are the stimuli for the affective responses,
not the direct musical stimuli themselves, unless the latter are
intellectualized and are exclusively musical, in which case 'the
affective experience will be similar to the form of the musical
form which brought it into being'.[40] For the ordinary listener,
however, there may be no necessary relation of the emotion to
the form and content of the musical work since 'the real stimulus is not the progressive unfolding of the musical structure
but the subjective content of the listener's mind'.[41]

Sadly, several of the surveys on the nature and frequency of
crying simply do not touch on the question of aesthetics. Fre-
derick Lund, who subscribed to the cathartic value of crying,
considered that aesthetically evoked tears required the presentation of an unattainable ideal. He suggested with regards
to music that the minor scale is more effective at releasing
tears than the major scale as it has the desired, combined with
the undesired, tonal elements falling short of consonance,
which arouses unrest and mingled feelings. Tears typically
occur when a sad or unpleasant setting is followed by a
redeeming element.[42]

In an unpublished work reported by Vingerhoets, over 100
students were approached with a 33-item questionnaire to
investigate the circumstances of crying. Three principal areas
were identified, namely aesthetics (poems, nature, songs, art,
etc.), films, and sentiments, for example in relation to pets and
social-related events. In a reflection of the habits of our era, it
was found that most tears are shed while watching the

television or at the movies. With regards to the arts, however, there was a difference between the different art forms in terms of their ability to release tears. Music rated highest on the list.[43]

When I lecture on the topic of crying, and its neurobiological basis, I usually begin by asking for a show of hands in answer to questions of who cries in response to works of art. I start with music, and inevitably about 90 percent of the audience will indicate that music makes them cry. I then go on to ask about other arts. For poetry, it is about 50 percent, but for painting, sculpture, and architecture the hand count drops towards zero. Music, it seems, is the art form that most readily captures the emotions, destroys composure, and binds listeners in communal rapture. Music moves us, and frequently moves us to tears: the word e-*motion* is six-sevenths motion, and there is a response to harmonic motion, the tonal system embedded with primal meaning—rising and falling, tension and rest, life and death.[44] Chills down the spine and crying are commonly recorded responses to music, increasing with the emotional intensity of the music, but correlating poorly with physiological measures of arousal. They do, however, correlate with changes within the brain, as we shall see.

John Sloboda studied the physical reactions of 83 people listening to music, and reported that 90 percent experienced shivers and 85 percent tears. The frequency of a physical response to a piece of music was found not to decline with repeated hearings, and tears or a lump in the throat were most often elicited by passages containing melodic appoggiaturas or sequences and harmonic movements through the cycle of

fifths to the tonic. Meyer predicted that satisfying music nego-
tiates between the expected and the unexpected, and Sloboda's subjects chose pieces which, in the words of Philip Ball,
'involved the kind of manipulation of expectation that Meyer
would have forecast, such as an acceleration towards a cadence
or a delay in reaching it, the emergence of a new harmony,
sudden changes of dynamics or texture, and rhythmic synco-
pation.'[45] Sloboda gives as a prototypical example of a tear
jerker the opening six bars of the third movement of Rach-
maninov's Second Symphony. Others high on the list were the
opening eight bars of Albinoni's *Adagio*, and the opening chorus of Bach's *St Matthew Passion*. The opening melodic line of
Albinoni's *Adagio* contains three appoggiaturas in the first
seven notes. When it comes to opera, sections from Mozart's
Cosi fan tutte, Puccini's *La bohème*, and Richard Strauss's *Der
Rosenkavalier* were top of the list.[46] Nietzsche said he was unable to differentiate tears from music.[47] It is said that the deaf
Beethoven was reduced to tears by reading through the score
of the Cavatina from his own String Quartet, op. 130.

According to James Elkins, whose book *Pictures and Tears*
analysed people's responses to paintings, Mark Rothko was
the only 20th-century painter who accepted the notion that
people might cry on viewing his pictures. Rothko, apparently
deeply inspired by Nietzsche's *Birth of Tragedy*, equated this
response to a religious experience. Others have used the word
'tragic' with reference to his paintings—tragic in the knowledge of his suicide in 1970, by an overdose compounded with
severance of the arteries of his arms; tragic in their emptiness;

tragic with tombstone-like despair? Whatever, tears in front of a Rothko seem to be his gift to only a few.[48]

Our art galleries are filled with tragic images, and much of the tragedy is associated with religious ecstasy or empathy, but Elkins himself notes that although he can be emotionally affected by such images, he never cries, and the emotional effect seems, for him at least, to dwindle with repeat exposure, in contrast to the reported effect of music. He was discouraged by art historians from even writing about the subject of crying, which in their opinion was irrelevant to the artistic process. The painter Oskar Kokoschka may have once cried in front of a painting; William Hazlitt admitted to crying on seeing Ludovico Carracci's *Susannah and the Elders*; and the composer Richard Strauss was moved to tears by Raphael's painting of *St Cecilia*, but crying on looking at paintings was not the experience of Leonardo da Vinci, the art historian Ernst Gombrich, or the large majority of the audiences at my lectures mentioned above.

Elkins's own experience is that, in contrast to paintings, films, and concerts, operas and novels were the commonest cause of crying as an aesthetic response.

After my lectures, several people have asked me why I did not include the cinema in my enquiries about crying in response to various art forms. In a study of crying at the movies, Suzanne Choti and her colleagues examined the responses of pairs of college students to their viewing of *The Champ* (1979) with a variety of mood and crying questionnaires.[49] Women were more affected, crying more often than men, and this was linked to measures of 'femininity' and the number of recent stresses in their lives. Men and women reported their expected stereotypes when in the presence of a man; in other words, women would respond with more crying and men with less if they were watching the movie with a man. For men, higher measured levels of tension, depression, and sadness prior to viewing were positively correlated with crying, but not for women. Further, the changes in scores for feelings between pre- and post-film ratings were different between the sexes. In women increased sadness, depression, and muscle tension were linked with crying, but not in men, who revealed decreased crying to be associated with increased anger.

From the perspective of my enquiries, the difficulty with cinema is that in films the dialogue is always accompanied by music, which in silent movies even portrayed the action and emotion. Weeping in the cinema results from the combination of the story, the music, and who else is present—imagine how many people would have gone to see Titanic (1997) without the music. The award-winning film The Artist (2012), written and directed by Michel Hazanavicius, was hailed as a silent film, but it was anything but that. As with the silent films of the past, the action and the emotion, and the aroused tears in the audience, were driven and orchestrated by the accompanying music.

Elkins, in trying to fathom why people may cry when they view paintings, concerns himself with religion. Tears are a part of religious iconography, representing sincere faith, and the weeping of religious statues is a part of the public paraphernalia of some religious cults, private crying being an offering to God. It is certain that the tales of the Gospels, which formed the content of many icons and paintings, evoked tears from those who dwelled on them and knelt before them in the Middle Ages and early Renaissance, but perhaps much less today. Elkins comments that in the 20th century, when it comes to viewing art, too much emphasis has been placed on teaching people to look and not to feel. In a modern art gallery, with its block buster exhibitions, we are surrounded by too many other people and distractions to do more than glance at the paintings, a 'bye-bye' view rather than a beholding. Further, with our contemporary distancing from religion and the historical relevance of religious art, we find it hard to contemplate them in the same way as in the past. Very often, the images have been torn from their original settings—namely religious dwellings, churches, and cathedrals—and are emptied of depth and devotion, being viewed superficially by a sea of spectators. Much contemporary painting is likewise etiolated, drained of its overt emotional potential but revived perhaps for the odd person when standing in front of a Rothko multiform or a Barnett Newman 'zip'. In a disturbing moment Elkins writes:

> At the eleventh hour, when this book was about to go to press, an Irish art historian, Rosemarie Mulcahy came up with an explanation as simple and as wonderful as it is devastating. Perhaps, she said, painting is simply weaker than the other arts, so it can't move people as music, poetry, architecture or the movies do.[50]

He hoped this wasn't true, but maybe she was right!

3

THE NEUROANATOMY
AND NEUROPHYSIOLOGY
OF CRYING

INTRODUCTION

The emotional responses to aesthetic pleasure, including crying, are of special concern to this book. From the surveys already reviewed, among the arts, music would seem to hold pride of place as the artistic form which most arouses emotions and provokes tears; less studied or noted is that novels are not far behind. The possible reasons for the prime place for music will be explored in Chapter 7, but for the present it is important to bring to the discussion what is known of the neurobiology of crying, since it is the purpose of the book to try to understand why crying emotionally is a unique human attribute, and why that is often a pleasurable experience.

WHAT DO WE KNOW ABOUT TEARS?

Tears are composed of water, mucin, proteins, ions, glucose, enzymes, lipids, urea, and a number of other chemicals. They have an important biological function, namely to keep the eyeball moist, especially to prevent drying of the cornea (which could lead to blindness) and to ward off infection. Blinking ensures the spread of the fluid across the eye, and when the eyeball is irritated, the fluid output from the lachrymal glands, which secrete tears, increases. In the human eye, there are many such glands: one main one and several subsidiary ones scattered around the conjunctiva, the transparent membrane that provides a protective cover to the eyes and eyelids. With normal basal secretion, fluid is drained via lachrymal ducts into the lachrymal sac which outlets through the nose. When the flow is excessive, the outlet through the nose is insufficient and tears overflow. Apart from the continuous production of tears which protects the eye, and the increased output with irritation, tears are released in response to emotion.

There have been studies showing chemical differences between tears induced by an irritant and those released by emotion. Earlier authors noted that emotional tears had a higher protein content. Frey confirmed this finding by inducing tears in volunteers with an onion and comparing their chemistry with those collected while watching a weep-inducing movie. He noted that the volume of tears shed was greater with emotional release and that there were differences in the content of peptides such as adrenocorticotrophin (ACTH), prolactin, and enkephalin between the tears.[1] Frey thought that, since ACTH is an indicator of stress, suppressing tears could increase stress and associated medical problems.

When tears from different species are compared, it has been found that the enzyme lysozyme is present in higher quantities in primate tears, and human tears are unique in having high concentrations of lactoferrin, an iron-binding protein, thought to be bactericidal. The primate with the nearest tear chemical profile to humans is the chimpanzee.[2]

It has been suggested that bodily secretions and excretions contain pheromones, substances with subtle smells, perhaps undetectable to consciousness, which have a powerful effect on behaviour, especially sexual behaviour.

No attention has been given to such properties in tears, but a recent study, in which males were given human female tears to sniff showed that they decreased sexual arousal in males, and lowered levels of testosterone.[3] How this relates to behaviour in the cinema is unclear, but within the context of mother–infant interactions a pheromonal effect of the infant's tears on bonding is worthy of study. As my colleague Mark George pointed out to me, of all the body secretions and excretions tears evoke the least disgust (personal communication).

THE NEUROLOGY AND NEUROANATOMY OF CRYING

To understand the neuroanatomy of crying requires some knowledge of the cranial nerves,[4] but even more about the representation of emotions in the brain, and the interplay of various neuroanatomical circuits in the regulation of lachrymation. It may be argued that understanding neuroanatomy is superfluous to an exploration of crying, as some would argue that it is also irrelevant for those interested in human emotions to study the brain—they will remain content to ignore the fundamental biology of their disciplines. This view should be relegated to history, evoking a time in the late 19th century when our understanding of neuroscience remained quite basic, and the two disciplines of neurology and psychiatry divided along lines which proved intellectually unsustainable, and which have been redirected and realigned by the explosion of findings in neuroscience only in the past fifty years.[5] A starting point in the search for an answer to the question of why emotional crying is confined among living species to *Homo sapiens* is neuroanatomy, for if distinguishing features of the human brain can be identified which provide an underpinning to the behaviour in question, the proposition gains biological plausibility, from which other conclusions, even about Tragedy, may be derived.

What follows will be a gentle guide through the brain, with comments on the historical developments of ideas in neuroanatomy, and then discussion of the most relevant structures for crying. A précis of the following pages is to be found in

Appendix 1, and those who wish for a gentler introduction may wish to read that section first.[6]

Anatomy of emotion

Anyone interested in this area is likely to be aware of the term 'the limbic system', and many have heard of the autonomic nervous system, with its sympathetic and parasympathetic divisions. Such terms are regularly used in newspaper articles and books which aim to provide a bridge between a topic of neuroscience interest and a social behaviour.

This is not the place to give an elaborate description of neuroanatomy, but it is the place to outline how knowledge of neuroanatomy has developed in the past few decades, and how an understanding of the new anatomy allows for a reinterpretation of the links between the brain and emotion which eluded previous generations. Most of what follows has been unravelled by careful clinical and basic neuroscience observations, but given much credence and confirmation in recent years with the use of new brain-imaging techniques. These are not described here, but capturing images of the brain that are highly resolved in time and space (milliseconds and millimetres) is now possible with such well-refined methods as positron emission tomography (PET) scanning, magnetic resonance imaging (MRI) and its variant functional MRI (ƒMRI), and magnetoencephalography. These methods, used in conjunction with the performance of a specific task, can allow hypotheses about specific associations between brain and behaviour to be tested.[7]

The neurology of crying

Crying is not only associated with an alteration of facial expression accompanying tears, but involves the respiratory muscles, with brief cycles of expiration and inspiration, changes in the tension of the vocal cords, and an outplay of the autonomic nervous system. Some have suggested that there is a crying 'centre' in the brain, and one approach to understanding brain–behaviour relationships is through the study of neurological diseases. In the 1920s the neurologist Samuel Kinnier Wilson (1878–1937) examined extensive clinical material in patients with pathological laughing and crying, and reviewed the literature on putative centres of emotion, although he failed to identify one himself.[8]

In a number of neurological conditions, control over emotion is diminished, and some form of pathological release of emotion may be seen. The term 'pseudobulbar palsy' is used to describe the easy flow of tears with the slightest sadness, as well as its opposite, disinhibited laughter at the mildest amusement. Interestingly, patients with such conditions usually do not have the underlying emotional feelings associated with the laughter or crying, and the problem is often of more concern to those who know and care for them, rather than to the patients themselves.[9]

The concept of 'centres' in the brain for certain functions is now quite outmoded, but the essential neurology remains intact. Thus pathological crying is seen in various disorders that affect the cerebral cortex, the larger area on the surface of the brain composed of nerve cells and their connecting tracts which descend and interconnect with other cortical areas or nerve collections (nuclei) lower down the brain's axis. These conditions include cerebrovascular disease and strokes, amyotrophic lateral sclerosis,[10] some dementias, multiple sclerosis, and diseases that damage some of those lower structures, such as the hypothalamus and pons, described in more detail below. Unfortunately, in many settings the lesions described are widespread, making any localization or even lateralization of the effect problematic. However, it is usually seen following lesions or disease of the frontal or the temporal lobes of the brain—divisions described further below—and the long nerve tracts, which carry information from the cortex of the brain to the nuclei in the midbrain and brainstem which control emotional expression. These, as will become apparent, play an important role in the neuroanatomy of human crying.

The evidence from epilepsy suggests that crying is related to circuits in the brain that underlie emotion. Thus, there are specific seizures, albeit not common, during which patients, as a manifestation of the attack, either laugh (referred to as gelastic seizures), or cry (dacrystic seizures). Dacrystic attacks are reported much less frequently than gelastic outbursts, and they have different anatomical associations. The site of the origin of the epileptic seizure in gelastic attacks involves parts of the cerebral cortex, especially those in the frontal and temporal areas and the hypothalamus (see below).[11] The neurologist Orrin Devinsky and his colleagues identified only eleven cases of dacrystic episodes in the literature, and reported on seven of their own. The anatomical site was most often in the temporal lobes, and in the majority of cases was right-sided.[12]

The beginnings of an emotional brain

Despite considerable interest in neuroanatomy and neuro-physiology in the 19th century, little progress was made in understanding how emotions were represented in the brain. At the end of the nineteenth century, the James–Lange hypothesis became popular. The physician and psychologist William James (1842–1910) suggested that the emotions were derived from sensory inputs to the brain, which activate motor outputs, and that the resulting bodily sensations are perceived as the emotion. However, there was no obvious cerebral location for the generation of emotion, although the sensory experiences were known to be received cortically, namely in the parietal cortex of the brain.[13]

James summed up his views on experiencing emotion as 'the bodily changes following directly the perception of the exciting fact, and…our feeling of the same changes as they occur *is* the emotion'.[14] In other words, we feel sad because we cry and not vice versa. The bodily changes are immediate and felt the moment they occur, and James considered it not possible to have an emotion without the associated bodily feelings. According to this theory, there is no selective central neuroanatomy of the emotions.

The Jamesian hypothesis was soon tested and shown to be wrong from two perspectives. First, it was shown that removal of the cortex of the brain on both sides in animals did not abolish the expression of emotion. Further, it was revealed that stimulation of various structures buried deep within the brain (beneath the cortex –referred to as subcortical) could lead to the release of emotion. These observations formed the basis for a revolution in neuroscience, the impact of which is still poorly appreciated, not only by many in the scientific community, but also by the public generally. It was the beginning of an understanding of the complex neurobiology of human emotion, and hence the discovery of brain circuits and neurotransmitters which interplay in the expression of emotion and control of feelings which, among other things, brought clinical psychiatry back from the wilderness.

The unravelling of the cerebral mysteries of our emotional being has been one of the most fascinating neuroscience endeavours of the last 100 years. We now appreciate that certain brain structures and pathways are crucial for the mediation and experience of emotion, and these are parts of our old evolutionary inheritance, which developed long before *Homo* developed into *sapiens*. The concept related to what is referred to as the limbic system. However, before describing the limbic structures and their functions, a brief orientation of some key brain structures is warranted.

THE BASIC COMPONENTS

The brain is composed of neurons, nerve cells which through a sophisticated machinery of ions, enzymes, and neurotransmitters, carry signals which drive the system. There are many other supporting structures, which include blood vessels, and the glial cells. The function of glial cells is still being researched, but in addition to providing structural support for neurons, they have metabolic and even neurotransmission properties. Neurons have cell bodies, axons, synapses, and dendrites as their main constituents (see Figure 1). The axon is a long thin tube which extends from the cell body to a distant site where it will synapse with other neurons. Within the neuron and along the axon is found a host of biochemical machinery and filamentous supporting structures, most of the relevant information for neurotransmission passing from the cell and its nucleus along the axon to the synapse. At the synapse, neurotransmitters are released which cross the synaptic cleft—a very small gap between one synapse and another—and exert a change in the electrical properties of the membranes on the other side of the synaptic cleft (post-synaptic membranes), which by and large are located on the dendrites of the post-synaptic neuron. These dendrites are small filamentous outgrowths of the neuron, the main points of information exchange in the neuronal system. Most neurons have a single axon, some branches, but many dendrites.

The number of neurons in the human brain is estimated in the billions, and the number of synaptic contacts in the trillions. Thus the number of potential states within the human brain is incalculable.[15]

The central nervous system is composed of several identifiable components: the cerebral cortex, the cerebellum (a smaller lobe at the base of the cerebrum), the basal ganglia,

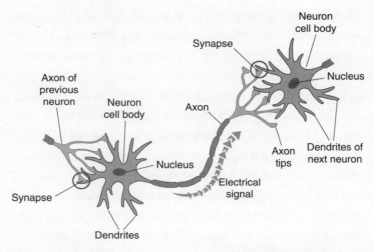

FIG. 1 The basic structure of the neuron, showing the cell body, axon, and synapse.

the brainstem (in which are found several nuclei that help reg-ulate emotional responses), and the spinal cord (see Figure 2). Information flows up the spinal cord from peripheral sensory receptors, which send signals to the brain about the external and the internal environments of the organism; there are many such *afferent* nerve tracts that have been identified within the spinal cord. Information travels to the muscles from the brain down the spinal cord, again in identified path-ways referred to as *efferent*. In the brainstem reside many of the nerves which control autonomic functions such as breath-ing, heart rate, and so on, and several of the cranial nerves including those linked with crying begin or terminate there.

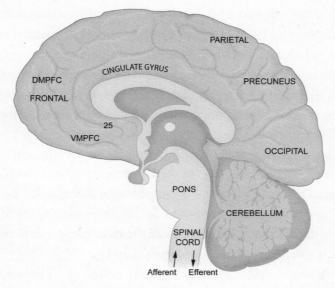

FIG. 2 The inflow and outflow of information up and down the spinal cord. Many nerve tracts synapse in the pons. Other important structures referred to in the text are also shown. DMPFC = dorsomedial prefontal cortex; VMPFC= ventromedial prefrontal cortex; 25= a part of the cingulate gyrus situated below the corpus callossum. The temporal lobes are not shown in this diagram.

THE BRAIN AND ITS DIVISIONS

The brain itself has been divided, purely on visual anatomical grounds into four main lobes, the frontal, the parietal, the occipital, and the temporal; some also refer to the insula component as a lobe. The cerebral cortex is usually referred to as the neocortex, so called because in phylogenetic (evolutionary) terms it is more recent.[16] The gyrae and sulcae form the irregular, undulating patterns of the cortical mantle, the gyrae being composed of grey and white matter and the sulcae forming the spaces between the gyrae. The grey matter is formed by the neurons, and the white matter is composed of the fibre bundles of axons, which stretch from one neuron population to another, interconnecting circuits of information.

Most sensory data initially terminate in a collection of nuclei situated subcortically, in a structure called the thalamus. This is one of several identifiable subcortical collections of nuclei which are important for an understanding of the emotional brain.[17] Others include the basal ganglia (referred to as striatal structures) (see Figures 3 and 4).

From the thalamus, information is passed up to the neocortex, arriving in a primary sensory area, such as one selec-

FIG. 3 The main structures of the limbic system: medial view.

FIG. 4 Connections between the frontal cortex, ventral striatum, globus palladus, and thalamus, with direct connections back to the frontal cortex. Note especially the input to the ventral striatum from the medial temporal structures, especially the amygdala. OFC = orbital frontal cortex; VS = ventral striatum; GP = globus palladus; Thal. = Thalamus; AMG = amygdala; STG and ITG = superior and inferior temporal lobe gyri; CD = caudate nucleus.

tive for vision or for touch. These are mainly situated in the posterior parts of the brain, in the occipital and parietal areas. There then follows a cascade of information, flowing from these primary receptive areas to the secondary, tertiary, and then to what are referred to as the association cortices of the brain. During this procession, the sensory representations are fused, amalgamated, and combined, so that while artificial stimulation of, say, the first visual receptive area will lead to an experience of flashes of light, stimulation in the temporal lobe association areas will lead to complex visual (and other) hallucinations. An exception to the above generalization is the olfactory system, which first enters the brain at one of the limbic nuclei and bypasses the thalamus[18] (see Figure 5).

The information flow out of the brain descends from the motor areas of the neocortex, down the pyramidal tracts to the spinal cord (called the pyramidal system), where the neurons synapse with other neurons which emerge from the cord or the brainstem to influence movements by connecting with

FIG. 5 Information flow from the primary sensory areas to the temporallobes and the limbic system. All sensory inputs except smell go initially to the thalamus. The two-way traffic shown is for vision to and from the amygdala, but it is the same for other modalities. Also shown are the outputs from the amygdala to the orbital frontal cortex (OFC).

muscle cells. Other key structures which influence movement and permit the effective carrying out of actions are the cerebellum and the basal ganglia, the motor paths of the latter being referred to as the extrapyramidal motor system.

The basal ganglia are a collection of subcortical nuclei which are extensively interconnected and which are related to cognitive and emotional behaviour. It used to be thought that these structures were purely motor in function, and that they had nothing to do with emotion or cognition. However, the shift in emphasis of their functions has provided neuroanatomical explanations for an understanding of actions—movement and being moved—that are the most important components of emotion.

THE CEREBRAL CORTEX AND THE LIMBIC SYSTEM

The key structures of the limbic system are the amygdala and the hippocampus, and their immediate connections (see Figures 3, 5, and 6). These include the orbital part of the frontal cortex (situated just above the eye sockets), the insula, and that part of the basal ganglia called the ventral striatum, which plays a central role in emotional–motor expression.[19]

The amygdala, shaped like an almond, is located at the front of the temporal lobes, and is central to the brain's regulation of emotion. It has extensive two-way connections with the neocortex, from which it receives combined (multi) sensory information, and to which it provides affective valence or tone to sensory information. The amygdala is the key to understanding the emotional resonances of our memories

FIG. 6   Some limbic structures seen from below, with part of the cortex cut away to reveal the underlying insula.

FIG. 7 A coronal section of the brain (cut from side to side) showing the parts of the medial and orbital frontal cortex. Connections between lateral prefrontal cortex, and the insula, with medial and orbital areas, are shown.

and, as will become clear, is going to play an important part in the unravelling of those emotional responses to music already discussed.

The hippocampus is also situated in the temporal lobe, and the cortex covering it (part of which is called the parahippoc-ampal-gyrus) also conveys complex integrated sensory infor-mation to the hippocampus from the neocortex (Figure 3). The hippocampus is closely involved with the laying down of everyday memories, which are given emotional tone by inte-gration with activity of the amygdala. One part of the basal ganglia, referred to as the ventral striatum, receives massive inputs from the amygdala and the hippocampus and, via a series of connections, tunes the neocortex to emotional events (Figure 4). Thus via the ventral striatum, emotion (limbic) is translated into motion (basal ganglia) and, with hippocampal input, the final motor output is finely primed to reflect past experiences, emotional states, and present environmental input from the senses.

Two important cortical structures for elaborating our emotions are the cingulate gyrus and the insula (Figures 3,6). The cingulate gyrus is an extensive structure which surrounds the brainstem, forming a C-shaped band, linking posteriorly with the hippocampal structures; anteriorly it melds with the medial frontal cortex, and throughout it connects extensively with neocortical structures. However, its widespread connections with the hippocampus, the amygdala, the basal ganglia, and other subcortical structures underline its importance for attention, motivation, and emotion.

The insula is a large limbic structure, which, in contrast to most limbic structures, is not visible from the medial surface of the brain, and lies laterally, but buried beneath the folds of the neocortex. It too has many functions, including integration of limbic and cortical information. It also links with the frontal cortex anteriorly, with the hippocampal structures posteriorly, and is the part of the limbic cortex which receives visceral information from the interior of the body. It is through such afferent activity that 'gut' feelings arise—the insula is a crucial structure in the circuitry of human emotion and crying.

THE FRONTAL LOBES

The frontal lobes of the brain have many demarcated subregions, but the orbital, medial, and lateral are the most frequently referred to. In neuroanatomy it is customary to refer to various parts of the cerebral cortex by a Brodmann number.[20] The parts of the frontal cortex most anteriorly situated, which relate to much human social and emotional behaviour are called prefrontal regions. They cover the lateral, medial, and orbital surfaces of the brain, and are extensively interconnected. The medial cortex includes certain areas that are closely related to clinical depression, and one of these, referred to as the subgenual cingulate area (or area 25: see Figure 2), has become a target of interest for deep brain stimulation in the treatment of chronic depression, but which, in animal lesion studies, has been shown to play a role in inhibiting autonomic behaviours. In humans this area projects to brainstem nuclei that regulate autonomic output, and is activated during the recognition of faces with emotional expressions.

The dorsolateral frontal cortex supports working memory, the suspension of information over time, to allow, for example, the tension of a poem or a piece of music to be held 'on line', and is associated with planning action sequences. It is more closely linked with the hippocampus than the ventromedial frontal cortex, the latter having much stronger insula and amygdala connections (Figure 8). Since the frontal areas mutually influence each other through intrinsic connections, so that coordinated motor plans emerge, these systems exert executive (top-down) control over the emotions.

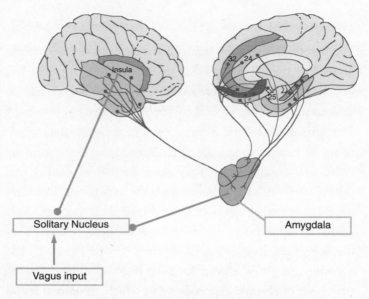

FIG. 8   A diagramme showing links between the amygdala, insula, cingulate gyrus, and frontal areas. It also shows the input from the vagus nerve
to the solitary nucleus, and the direct pathways from the latter to the insula
and amygdala. This reveals how 'gut feelings' get into limbic structures. Not
shown are the inputs to the subgenual cingulate (area 25) from the insula.
The numbers 25, 32, and 24 are Brodmann numbers.

Laterality, namely which side of the brain is especially involved
in an activity, is important, since the right ventrolateral prefrontal
cortex has been shown to be related to the recall of autobiographical memories which are so important in the evocation of tears.

The orbito-frontal cortex has intimate connections with
the insula, the amygdala, the basal ganglia, and the hypothalamus but importantly, in the human brain, this and several
other frontal regions project further down the neuroaxis to

the brainstem. It is through these pathways, the anatomy of which is different in the human brain from other primates, that frontal activity can directly influence and control our emotions by altering the activity of those brainstem neural groups that finely temper our emotional state.

Towards a more complete understanding of the neurology of emotion

As we have already seen, James could not satisfactorily explain how emotions could be represented in the brain. The concept that certain brain structures, referred to as limbic, could form the foundation of an emotional brain system was a stunning departure for neurology, and a revelation as profound to human science as the discoveries of Copernicus, Darwin, or Freud. The key pioneers in unravelling this neuroanatomy, one of the most exciting neurological expeditions of all time, were the anatomists Walle Nauta (1916–94), Paul MacLean (1913–2007), and Lennart Heimer (1930–2007), and their numerous collaborators.[21] In the mid twentieth century they challenged the belief that cortical and subcortical systems were distinctly separated, and noted the strong connectivity between limbic structures, the basal ganglia, and the neocortex, therefore overturning entirely the idea that the limbic system was a discrete system regulating emotion, which was unable to influence the basal ganglia (motor systems) or the neocortex (and thus speaking, doing, and knowing).

These researches also revealed the extended influence of the limbic system to the midbrain (some refer to a limbic midbrain) and brainstem, including connecting with the cell structures that we now know are the origin of the ascending neurochemical systems, especially of those neurotransmitters called monoamines. These transmitters, such as dopamine, serotonin, and noradrenaline, widely influence emotion and motor activity by projecting back up from the lower centres to the basal ganglia, limbic structures, and the neocortex. It can only lead to wonder that these transmitters, which arise from very discrete nuclei in the brainstem, influence so much of not only human but also all vertebrate behaviour, and to fascination that they are involved in so many human neuropsychiatric disorders from Parkinson's disease to schizophrenia. From an evolutionary standpoint these neurochemicals are very old, and have been a driving force of evolution. No dopamine, no movement; no serotonin, no emotional tone; no drive, no survival.[22] The limbic outflow also affects the neurons of the autonomic system and the cranial nerves that innervate the muscles of facial expression—without which we cannot smile, laugh, or cry.

To summarize so far: The inputs to the limbic structures have been shown to be both interoceptive (from within the body) and exteroceptive (conveying information about the immediate environment). The former derive from many bodily structures (muscles, blood vessels, the viscera, etc.) that give information about the internal state of the organism, and include modulating influences from those neurotransmitter pathways that help drive behaviour and modulate mood. The exteroceptive data derive from all sensory systems that reveal the environment of the organism, which are presented as complex integrated information to the hippocampus and the amygdala.

These limbic structures interact with other key ones such as the cingulate gyrus and the insula, which because of their access to information about the state of the viscera (guts), coordinate activity through the basal ganglia to influence motor responses, and autonomic activity via the hypothalamus. The latter structure also coordinates such actions as

FIG. 9   Sensory inputs from the solitary nucleus to the parietal cortex and insula. Also shown is the motor output to the structures that relate to crying from the other vagus nuclei. NA = nucleus ambiguus; DMN = dorsal motor nucleus of the vagus; NTS = solitary nucleus.

the release of hormones, which flow from glands such as the pituitary gland at the base of the brain and further influence our feelings by their effect at peripheral organs. The brain's regions that modulate emotion and motivation thus have direct access to the motor output systems of the brain, but their direct influence extends further, down as far as the brainstem to the neurons that control voluntary and autonomic muscular activity. The input side of these circuits is shown in Figure 9.

The autonomic systems

Reference has been made to autonomic activity, and now some more explanation is required. In contrast to the somatosensory and somatomotor systems, those nerves which provide us with information about the outside world and influence the parade of our essentially voluntary muscle activity, the autonomic system controls those bodily actions that help maintain the homeostasis of our internal economy. For example, regulation of heart rate, breathing patterns, or temperature control through sweating. Generally we do not consciously alter these functions, but we can do so, some by using techniques such as biofeedback, others by adopting autogenic means. The ability of each one of us to intervene in the activities of our autonomic nervous system varies.

The autonomic system is closely related to our emotional state, and is itself influenced by factors that alter our emotions. Our current understanding combines knowledge of the limbic structures and their extensive circuitry described above, and the theories of James, in recognizing the role of peripheral and visceral sensations in emotional feelings and as markers for action.

There are *sympathetic* and *parasympathetic* divisions of the autonomic nervous system, and their distributed anatomy is complicated. What matters here is to note that, while in the past the two were seen as opposites—for example the sympathetic division increasing heart rate and the parasympathetic decreasing it—the two are now thought to be more harmonious in action, maintaining our internal homeostasis within quite narrow limits necessary for minute-to-minute survival—cooperation, not competition.

Parasympathetic output is from the brainstem and from the lower spinal cord, although it is only the former that is of relevance here.[23] The brainstem visceral division comprises a number of cell groups adjacent to the central canal (or aqueduct—filled with cerebrospinal fluid, the liquor which also surrounds the brain and provides, among other things, protection during movement), the output of which, travelling with various cranial nerves, send neurons that supply the muscles of the head, thorax, and abdomen. Crucial to crying is the activity of the tenth cranial nerve, the vagus nerve, and its nuclear clusters where it originates from in the brainstem. This nerve provides innervations that control pupil size, salivation, and lachrymation (tearing), but it also outputs to many other internal organs, including the whole of the digestive tract (see Figure 9).

The nuclei linked with the vagus nerve form part of a carefully coordinated complex for regulating heart, lung, and gastrointestinal tract activity. At one time it was thought that the traffic of the vagus nerve was primarily outgoing, but it is now known that it carries much incoming information to the brain from the gut and other structures that it influences. These afferent impulses end in a collection of nuclei in the brainstem.[24]

The neuroscientist Stephen Porges has provided clues to the origins of emotional expression in *Homo sapiens* through an analysis of the control of autonomic, especially parasympathetic, outputs from an evolutionary perspective.[25] He describes how, in the transition from reptiles to mammals, there has been a huge development of what he refers to as the 'social engagement system', which relates in part to alteration of the neural circuitry of control of the facial muscles. He draws attention to an anatomical distinction between the neural outputs regulating supradiaphragmatic structures from those acting on infradiaphragmatic ones.[26] Supradiaphragmatic structures are governed by myelinated vagus fibres which stem from a vagus nucleus called the nucleus ambiguus, while infradiaphragmatic ones are unmelinated and come from another vagus nucleus, the dorsal motor nucleus of the vagus. Myelinated nerves are surrounded by a coating of myelin which speeds up transmission, thus allowing for rapid responses to events that trigger emotion. It is the nucleus ambiguus that is the most important in emotional expression.

In development, myelination continues especially in the first three months of life. Further, with ageing there is increased cortical influence over nuclei which not only control the muscles of the head and face, but also the brainstem nuclei which output to the heart and lungs, larynx and pharynx.[27] Our social communications involve coordinated activity of facial expression, heart rate, the larynx, pharynx, and breathing, and this is obvious in crying. The physical act of crying is mainly one of inspiration, but the anatomy also involves the soft palate, larynx, and pharynx. It is very destructive of human speech; with tearing we become choked, speechless. This suggests that emotional crying (and laughter—which occurs largely with expiration) evolved before propositional language, perhaps explaining why tears communicate states of mind that are often so difficult to express in words.

The nucleus ambiguus receives top-down neural information from limbic structures, via the insula and hypothalamus, but also from the frontal and prefrontal cortex including the cingulate area (see Figure 11 below). In addition to that, there is input about the internal state of the body via the other vagus nuclei, such as the one referred to as the solitary nucleus, which receives many incoming fibres from the viscera. The three vagus nuclei thus form a lattice of sensory–motor information integrating and coordinating autonomic activity.

The solitary nucleus projects directly up to the insula—the visceral or autonomic cortex—such that the feelings in the gut (and any resulting tears) reflect the interplay of afferent and efferent projections of the vagus nerve, the nuclei of which are under the influence of limbic and cortical projections from above. The parasympathetic fibres from the nucleus ambiguus and lachrymal nucleus travel via the lachrymal nerve to the lachrymal glands, as shown in Figure 10.[28]

Sensory input from the vagus nerve reaches the primary sensory cortex in the parietal lobes from the thalamus, and is relayed forwards to the insula and the frontal cortex (Figures 9 and 11). The feedbacks of this circuitry are clear, namely the neural traffic from the nuclei regulating autonomic activity in the brainstem go to the neocortex and the latter provides top-down regulation to the same nuclei, with several other structures such as the cingulate gyrus and the insula being involved

FIG. 10 The outflow from the brainstem lachrymal nucleus to the lachrymal glands via the cranial nerves. The two cranial nerves that carry this reflex are the fifth (trigeminal) and seventh (facial). Nucleus of solitary tract = Solitary nucleus

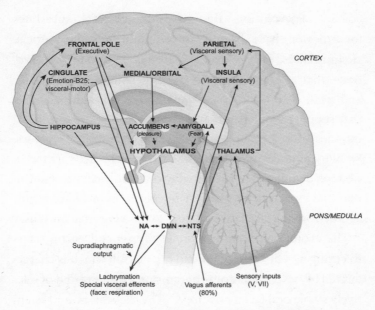

FIG. 11   Connectivity between the inflow of information from the vagus nerve nuclei and other sensory information to the limbic structures and frontal cortex and the outflow back down to the vagus nerve nuclei. Note the direct links between frontal and cingulate areas and the nuclei of the vagus nerve, which are a feature of the human brain.

in the loop. Thus the importance of higher cortical control in the regulation of human crying is clear neuroanatomically.[29]

But the emotional state of crying is not purely one of increased parasympathetic tone, since it is associated with considerable sympathetic activity. There is both a calming (parasympathetic) and a mobilization (sympathetic) and, in the course of evolution, the neurophysiological substrates for emotion have been crafted and moulded by social circumstances. Porges makes a further important observation—that afferent impulses from the muscles of the face and head play back to the same vagus autonomic nuclei that supply them, such that activation of the eye and facial muscles, as in looking or grimacing, trigger central changes. So why not crying? This theme is taken up further in Chapter 6, but this is presumably one neuronal mechanism that allows actors to turn on the tears, by a reflex motor action from the facial musculature to the lachrymal nucleus.

The evolutionary anthropologist Terrence Deacon refers to crying as a distinctive human innate call, which is characterized by 'visceral motor programs on a background of relatively stable oral facial postures'.[30] The skeletal motor system is subordinated to the visceral system, and this has a basis not only in terms of comparative anatomy, but also in the value of crying. The act of crying is, like most primate calls, closely related to activity in cingulate, midbrain, and brainstem structures and, along with laughter plays a fundamental role in social communication and empathy, evolving before the development of speech as we know it today. However, a vital part of the story lies with the symbolic nature of human speech and the ability of narrative to influence emotional states. This is in part dependent on prefrontal cortical activity, and linked with an ability to understand that others have minds like one's own, which is referred to as Theory of Mind and discussed in detail in the next chapter.

Deacon observes that 'the symbolic construction of others' plausible emotional states, and their likely emotional responses to our future actions, are analogous to a whole new sensory modality feeding into our ancient social-emotional response systems'.[31] As Porges suggests, the phylogenetic origin of the behaviours associated with the social engagement system is intertwined with the phylogeny of the autonomic nervous system.[32] The vocal cry of the infant becomes quiescent and controlled in adolescence, but the tearing remains throughout life, and the same anatomical patterns and facial contortions associated with it are seen across the ages and throughout the world.

HOMO SAPIENS

At this point I shall readdress some differences between primates and other animals and between *Homo sapiens* and related primates from a neuroanatomical perspective.

In the course of mammalian development, a number of fundamental shifts of brain structure have occurred, especially with the development of nuclei and tracts related to enhancing emotional expression. The limbic-related structures are not simply appendages to the developing reptilian axis, but are tightly integrated with those basal ganglia structures which *move* the organism, and with the developing neocortex. As we progress up the phylogeny towards mankind, the extravagant enhancement of the neocortex, reaching its pinnacle in *Homo sapiens*, may have afforded us untold evolutionary advantages, but there was no separation from limbic influences—emotion was still the driving force of behaviour. There was, however, a shift in the direct down-path cortical control over brainstem nuclei, allowing for top-down regulation of autonomic activity. The situation is summed up as follows, and illustrated in Figure 12.[33]

In what the neuroscientist Georg Striedter refers to as Deacon's rule, in evolution, when a brain region becomes disproportionately large, it tends to invade novel targets and receive novel inputs.[34] In mammals, and therefore in primates,

FIG. 12 Diagram showing increasing cortical control of top-down structures on brainstem nuclei regulating autonomic activity from the mammal outline (*left*) to the human (*right*). In the human, the output to the structures that regulate crying have many more direct connections, which is less in other primates in whom there is more relay in the pontine reticular premotor nuclei.

the motor output from the cortical areas controlling muscle activity in the spinal motor neurons project directly and more densely to those neurons as brain size increases, one reason for the increased motor skill of humans in comparison with other primates, with our very fine manual dexterity. This allows us to oppose the thumb with the index finger forming a circle—a universal gesture for perfect. No other primate can do that.[35] It is the same with control of the muscles of the face, tongue, and vocal cords, such that the direct cortical input influences vocality and social interaction which develops alongside increased flexibility and fecundity of facial gestures. Such top-down influences in Homo sapiens mean direct frontal cortical control over the nuclei regulating autonomic activity, and therefore crying. This allows for a more flexible organization of human emotional behaviour, as the coordinated physiological effects of crying and its induction by social circumstances become embedded in autobiographical memories.

A further observation is the relative change of size of the visual cortex in the human brain. Thus in other primates, the visual cortex correlates well with brain size, but in Homo sapiens it is smaller. One hypothesis is that this implies increases in size elsewhere in the brain, and Striedter notices, among other areas, the increased size in the temporal lobes, especially the dorsal area that relates to the auditory reception of speech—heralding a shift to an aesthetics based on sound.

There is now good agreement as to the areas of the brain involved in the generation of emotional states, and this has received verification from recent brain-imaging studies. Antonio Damasio and his group have examined in detail the cerebral events that follow on from emotional arousal. Their starting point is to distinguish emotion from feeling, a separation which James did not make. Emotion, as we have noted, entails motion: it relates to movement, to action, and develops from unlearned brain programmes of automatic actions and cognitive strategies, the purpose of which is to promote the organism's survival. Emotions are part of the regulation of homeostasis. From a neuroscience perspective, more complex emotions are built upon simpler ones, but use the same neuronal circuits, interlinked with evolutionarily guided reward and punishment mechanisms.

Emotions thus have biological value. Arguing from observations of the changed behaviours of patients with ventral and medial prefrontal brain lesions, Damasio's somatic marker hypothesis proposes that the stimuli that induce an emotion activate pre-existing neuronal circuits (initially in frontal cortex), triggering activity in both body and brain. These somatic marker associations relate to the present state of the body, even if they do not arise in the actual body but are in the representations of the body in the brain. They serve to bias responses, either consciously or unconsciously, to the environmental situation.

There are various ways that the emotions have been classified, but Damasio suggests there are primary emotions, background emotions, and social emotions. The first embrace fear, anger, happiness, sadness, and disgust. The background ones are such emotions as enthusiasm and discouragement, while the social emotions are shame, guilt, contentment, compassion, jealousy, awe, and revenge. These emerged at a later stage in evolution than the primary emotions, but several, for example compassion, have been shown not to be exclusively human. They obviously shape, and are also themselves shaped by, cultural factors. These emotions provide an attitude and a protension towards the world, but action is the crucial ingredient.

Emotions are triggered by external and internal events. Damasio refers to these as emotionally competent stimuli, which act on cerebral circuits which have been present within the structure of brains, and thus within the human brain, for millions of years. Some of the extrinsic stimuli and the neuronal events they evoke have been embedded in behaviour much longer than others. Fear is very old in evolutionary terms, while the stimuli, especially those associated with the social emotions, may be learned and the responses more protean. Different emotions and different circumstances are tuned with different memories, attention, and expectations.

Emotions are thus felt either as changes in the bodily state or as a particular pattern of body representations in the brain, the latter being a learned reinstated state from past memories—an 'as if' pattern generated within the brain. Damasio's hypotheses are shown in Table 1. In the context of this book, crying is seen as an emotionally competent stimulus, having an effect on the individual who is crying (somatic), but leading to a response in another, which is evolutionarily embedded and automatic.[36]

TABLE 1. **Damasio's concept of emotion**

1.	An emotion-proper, such as happiness, sadness, embarrassment, or sympathy, is a complex collection of chemical and neural responses forming a distinctive pattern.
2.	The responses are produced by the normal brain when it detects an emotionally competent stimulus, the object or event whose presence, actual or in mental recall, triggers the emotion. The responses are automatic.
3.	The brain is prepared by evolution to respond to certain emotionally competent stimuli with specific repertoires of action. However, the list of emotionally competent stimuli is not confined to those prescribed by evolution; it includes many others learned in a lifetime of experience.
4.	The immediate result of these responses is a temporary change in the state of the body proper, and in the state of the brain structures that map the body and support thinking.
5.	The ultimate result of the responses, directly or indirectly, is the placement of the organism in circumstances conducive to survival and well-being.

Source: Damasio, *Looking for Spinoza*, p. 53.

The psychologist and neuroscientist Edmund Rolls has also placed an understanding of human emotions within a Darwinian context. Emotions—which he defines as states elicited by rewards and punishers, or what are called instrumental reinforcers—are closely allied to the activity of the amygdala and the orbito-frontal cortex.[37] The events felt as emotions are aroused from external and internal bodily stimuli, and conscious awareness is not initially involved. Porges uses the term 'neuroception' to describe this process, different from perception, which is linked to an appraisal of risk in any particular situation, but which allows social behaviours to proceed.[38] Cognitions can modify the responses to the stimuli, in part through the effects of language and memory, allowing for the conditions in which an emotion is experienced to be influenced by higher cognitive functions and evoked by cultural activities such as reading a novel or seeing Tragedy.

Using ƒMRI, Damasio and his colleagues have traced the brain changes seen immediately following the induction of an emotional state, showing how the neuronal structures discussed above perform a cascade of events over some 500 milli seconds following the provoking stimulus. In considering the anatomy in some detail, Damasio notes that the right insular cortex and areas of the right parietal cortex hold an integrated representation of the ongoing internal state of the organism, these regions of the parietal cortex in the right hemisphere mapping the body state, the insula representing the area of limbic cortex receiving visceral sensations.[39] These areas are activated, along with the temporal cortex and amygdala, by emotional stimuli. The prefrontal areas, associated with pre-vious emotional experiences and autobiographical memo-ries, provide social and personal contexts, leading to a flow of neural activity involving the cingulate gyrus, the orbito-frontal cortex, and their outflows to the brainstem structures which coordinate the autonomic responses. The gut feelings reverberate via the vagus inputs and their reiteration to sen-sory cortices and the insula.[40]

To summarize, in the human brain, at the top of the hierar-chy of the cerebral circuits for emotion is the higher-order neocortex, especially the sensory association cortex in the parietal lobes, and the medial and orbital prefrontal cortex. Fear is tightly bound up with the activity of the amygdala, while the social emotions involve especially the insula and the anterior cingulate. The subcortical structures involved are the basal ganglia, notably the ventral striatum, the hypothala-mus, and several brainstem structures including the vagus nuclear complex. It is important to re-emphasize that the flow of information is nearly always two-way, top-down and from the brainstem to the cortex, in a mutual harmony of reverberations which modulate the emotion, prolonging it, controlling it, or modifying it depending on circumstances.[41]

Tears

To weep we need tears. The main lachrymal gland is activated by the parasympathetic division of the autonomic nervous system, the seventh (facial) and then the fifth (trigeminal) cranial nerves carrying the nerve impulses to and from it, using the lachrymal branch of the ophthalmic division of the fifth cranial nerve (see Figure 10). Some texts refer to a poorly defined lachrymal nucleus in the brainstem, although its exact designation is quite unclear.[42] In the human brain this receives sensory innervations from the trigeminal nerve, which allow for reflex stimulation from the muscles of the face, but also input from the hypothalamus, thalamus, basal ganglia, and cerebral cortex, providing the top-down control already referred to.

Botulinum toxin

Recent neurological and neurophysiological evidence has shown that feedback from around the eyes and from the muscles of the face to the brain influences emotional states. Botulinum toxin injections have become a fashionable method for the removal of facial wrinkles. They were first introduced for the treatment of muscle disorders such as dystonia, and their effect is to weaken the strength of muscles, without necessarily paralysing them. According to the James–Lange hypothesis and the Damasio revision, feedback from the somatic as well as the visceral muscles contributes to the feeling of the emotion. Increasing facial expression has been shown to increase autonomic activity, and self-reports of experienced emotion, and mimicry of an emotion, can activate the same cortical areas as observing it, although the effect is greater with mimicry. Using botulinum injections to induce temporary denervation of those facial muscles associated with sadness, in combination with *f*MRI imaging, Andreas Hennenlotter and his colleagues have shown that the reduced feedback decreased the expected signal changes in the orbito-frontal cortex and brainstem.[43] Thus afferent input from the facial muscles and skin directly influences neuronal activity in circuits that are linked with emotion. Newborns have been shown to imitate facial gestures, suggesting innate brain mechanisms for this skill, and hence for the later release of tears via feedback from the muscles and skin around the eyes.

Tears are thus stimulated by irritation via a reflex along the trigeminal nerve, leading to a needed quick response. However, tears in response to emotion also involves direct input from the higher neuronal circuitry, and the cerebral structures that are bound to bonding, the social brain, to be described shortly.

Laterality

Although at first glance the two cerebral hemispheres may look quite symmetrical, on closer inspection they are not, and each subserves different aspects of the overall behavioural repertoire. In the course of evolution, the potential for lateralization, which has been shown in other species, has been further developed in the human brain, the result being that our brains have become lopsided.[44] There are good reasons for this, allowing for abilities that do not require bilateral representation in the brain, such as speech as opposed to sight, to be preferentially localized in one or other hemisphere. This minimizes brain size and energy consumption, and allows for continued safety of the process of birth in mammals. If the head size of the foetus were to grow too large, it would not be able to travel through the birth canal, which would be dangerous for the mother as well as the infant. Further lateralization of function allows neurons with the same type of computational activity to be placed together, aiding brain development and minimizing the need for long-distance connecting axons.

In neurology and in human neuroscience, most attention has been given to the left hemisphere, which, in most right-handed people regulates syntax, propositional language, and the motor abilities of the right side of the body. It is referred to as dominant, in contrast to its non-dominant counterpart, the right hemisphere. Functional asymmetry of the brain in humans begins early in life: studies of cerebral blood flow show a right hemisphere predominance until about age 3, when a shift to the left occurs.

The right hemisphere has been known for some time to be associated with visuo-spatial functions, but it is only recently that the much wider perspective of that hemisphere has become appreciated and investigated.[45] The non-dominant hemisphere has had its champions, including the psychiatrist and philosopher John Cutting, and more recently Iain McGilchrist in his brilliant analysis of the interactions between the two hemispheres, *The Master and his Emissary: The Divided Brain and the Making of the Western World*. In contrast to the way the left hemisphere deals with the world it has to negotiate, the right hemisphere seems dominant for attention, for attending

to bodily space, for integrating circuits for spatial representation of the body, and, importantly, for regulation of mood
and certain aspects of speech, the prosody. The right hemisphere utilizes holistic, global representations rather than the
discrete, particularized boundaried computations of the
dominant hemisphere. According to McGilchrist, the right
hemisphere has virtually all of the attributes that make us
human. It is attracted to what is new in the world, is attentive
to the world around us before the left hemisphere, is the
repository of our emotional biographical memories, and is
where our embodied sense of self is located. The left hemisphere, in contrast, deals with only what it knows; syntax
bound, it constricts meanings that are only relevant for its
own achievement. The right hemisphere comprehends metaphor, is creative and insightful in problem solving; it is deduc-
tive and responds to the minor key.[46]

What is less clear is whether or not the older limbic and
subcortical structures of the brain also have lateralized functions, and if they do how this may influence, or in the course
of evolution, have influenced, the overlying cortical development. It is most likely that this is the case, since some suggestion of lateralization of function can be found in species all
the way along the phylogeny from fish to non-human primates, which relates not only to basic behaviours, but also to
everyday social repertoires. There is good evidence from primates, and from brain stimulation or other neurosurgical
intervention studies in humans, of limbic lateralization. It has
been shown that, for example in rhesus monkeys and mar-

mosets, there is a population bias for the right hemisphere to control emotional expressions. In humans, the right amygdala is associated with fear and panic. Tachycardia recorded during epileptic seizures is associated with right temporal lobe epilepsy, and human micturition seems regulated by a group of cortical and subcortical structures predominantly right-sided. There is also evidence for lateralization of function in the human insula, especially with regards to the control of autonomic activity.

The two hemispheres thus seem to employ different processing mechanisms, and the so-called non-dominant hemisphere is truly dominant for some fundamental aspects of human emotional and social behaviour. Essentially it is the hemisphere that hears and speaks not the language of philosophers, but that of poets and musicians.

4

EVOLUTION

Where and how does it begin...? A question of origin.

Derrida, *Of Grammatology*[1]

The human species represents just one stage in the story of life on this planet, a process that has lasted some four billion years, during which time perhaps 99 per cent of species that have ever lived have vanished off the face of the Earth.[2] Mammals appeared about 50 million years ago, and the earliest hominids some four to four and a half million years before those of us who are here today.

There were the bipedal walking but not talking Australopithecines. *Homo habilis*, with a slightly larger brain, evolved some 2.5 million years ago, a maker of stone tools and user of fire. *Homo ergaster* followed—a more recognizable human form and user of more sophisticated tools. *Homo erectus*, who resembled modern humans, appeared 1.5 million years ago, and made tools for future use. They may have possessed some form of spoken communicative language. The largely European-dwelling *Homo neanderthalensis* (Neanderthal man) evolved some 300,000 years ago, and then, with *Homo sapiens*, perhaps 150,000 years later, arose the beginnings of human cultures as we know them today, with complex languages, religion, myth, music, Tragedy, and tragedy.[3]

As we approach the era of modern humanity and discover more and more about our past, the dates for the emergence of hominid types and intellectual properties through evolutionary history seem, with new discoveries, to recede ever further back in time. In addition, clear distinctions between *Homo sapiens* and primate predecessors seem to dwindle. Understanding what is uniquely human is quite a game of catch me if you can—but in this book emotional crying has been selected as just such a trait, a marker of true humanity.[4]

LANGUAGE

Most people identify our ability to communicate, using spoken languages, as a uniquely human enterprise, but as yet we do not know how or exactly when this capacity arose. Language, in a more general sense, goes way beyond the spoken word, and encompasses acts such as gesture, pantomime, and mimicry. It has been speculated that some form of language ability arose with *Homo habilis*, but not one as we know it today. Various authors speculate about proto-languages. One, Stephen Mithen, refers to the musi-language of the singing Neanderthals, which he believes evolved from primitive primate calls.[5] The anthropologist and linguist Robbins Burling wrote: 'Primate calls have much less in common with human language than with human screams, sighs, sobs and laughter. Our own audible cries, howls, giggles and snorts, also with our visible scowls, smiles and stares, all belong to our primate heritage.'[6]

Thoughts must have preceded spoken language, in the sense that we cannot deny that other primates have 'thoughts' about their environment and about how to act effectively within it, even if those thoughts do not have the structure of human propositional language. There can be thoughts without words—though this proposition would be vigorously denied by many philosophers. The motor outputs we use for communication and call language are but a means of translating thoughts and interacting with others or the self, and any pre-linguistic analogues cannot resemble the sophistication reached by *Homo sapiens*. Nevertheless, according to Mithen, some early primate vocalizations not only sounded like human chatter but were rhythmical—it is well known that our remote predecessors, the mainly monogamous gibbons, sing duets.

The singular origins of spoken language and music have been suggested by many, including the philosopher Jean-Jacques Rousseau (1712–78), who put forward the view that it was passions rather than needs that provoked the first words.[7]

The chain of events links emotions to thoughts, to some preverbal intermediary, then to the spoken word, probably via gesture and indexication to spoken language. The importance of gesture for the origins of human speech has been strongly championed by authors from the French philosopher Étienne Bonnot de Condillac (1715–80) to Corballis.[8] Michael Corballis and Thomas Suddendorf note the relevant shift from a visual (gestural) to an auditory (speech) system that our languages require: 'The human equivalents of primate vocalizations are probably emotionally-based sounds like laughing, crying, grunting, or shrieking.'[9] With evolution, primate facial muscles became more flexible, facial gestures more elaborate, and social communication more complex. The 'social brain' expanded; what Porges refers to as the social engagement system engaged, and with it the power of facial expression to signal emotions. As a part of this, at a point on evolutionary time to be discussed later, tears became a cipher of singular importance.

The essential feature of spoken and written language is its symbolic nature. Words have no direct connections with the objects they denote. They refer to concepts, and such concepts are cerebral, and rely heavily on metaphor, which is central to the construction of meaning. When did the enhanced power of symbolic thinking erupt on this planet? Some postulate a 'big bang', an explosion of cognition, probably enhanced by genetic changes influencing brain structure and function some 40,000–50,000 years ago. Others favour a more gradual evolution. There is evidence of artefacts such as spears for throwing, figurines, and bone harpoons that date back perhaps 400,000 years. The use of pigments, per-haps used for decorating bodies, and the grindstones for mak-ing them, have been located in several archaeological sites, dating back at least 100,000 years. There is even some evi-dence that later Neanderthals developed cultural traditions with symbolic intent. If these later findings hold up, then symbolism, the driver of the culture of *Homo sapiens*, and the cerebral potential for artistic and cultural creativity are older than we think.[10]

MYTH AND RELIGION

Religious rituals have been traced back to at least the Nean-derthal time, about 300,000 years ago, but the dating of such epochal events is only speculation. Some kind of reli-gious ceremony may have been practiced for more than a million years. The after-world, or some kind of ancestor abode, surely was 'revealed' to the individual through dreams, in which the dead were reanimated, the past revived, and the future revealed. It is known that non-human pri-mates, and indeed many mammals, dream. The precise functions of dreaming are unknown, but even today sen-tient humans revere their content highly—surely they were a potent force in the development of some cultural events in early times?[11]

All spoken cultures have their myths of creation, psycho-logical symbolic narratives read as allegories or metaphors: our inner world transformations of the outer world. Myths frame the intuited order of nature for a group of people, and help individuals navigate their way through the stages of life from birth to death and perhaps beyond. The earliest myths had narrow horizons, were local and tribal, and were closely bound with nature and the natural environment of the group. Joseph Campbell refers to mythic 'universals', albeit locally and socially conditioned, which can be found if one looks for them.[12] For authors like Carl Jung (1875–1961) and Freud, these were innate, immanent, archetypal developments of the early human psyche, which somehow remained embedded within the structure of the evolving human brain.

The metaphors employed in any mythology may be defined as affect-laden signs and symbols, derived from intuitions of the self and the community. These become revealed through ritual, prayers, poems, meditations, ceremonies, annual festivals, and theatre. Out of the early individualized and collective myths emerged more formalized religion, but the feelings and cognitive structures that drove them must have preceded this development. Our languages distinguish us, not only from all other species of all evolutionary time, but embed our individual psychological expressions, and encase our social and cultural boundaries. Although we have no access to the ancient hominid mind, we can assume that it was fundamentally different from our own, especially with regards to language. What the situation was for, say, *Homo habilis* is a complete mystery.

Like everything in evolution, however, things could only have graduated; there was no sudden enlightenment of the world's primates, only a progressive questioning of the meaning of life and the rise of the individual consciousness. It is known that in Palaeolithic times, some 100,000 years ago, the dead were buried with objects, presumably to help them with a journey or to deal with some kind of after-life existence. Tools and hunting weapons, food and flowers, aided the preliterate Neanderthal on an ethereal way, to a world that, if not known, could be imagined. Later the dead were buried with ornaments, necklaces, and beads, and the early semblances of religious art are observed.

An early modern human with a high rounded forehead, inhabited Europe in the middle Stone Age, some 100,000 years ago.[13] *Homo*, by now *sapiens* and language-fluent, provided the first examples of prehistoric art and developed a complexity of spiritual beliefs, which included some idea of the gods.[14] Matt Ridley comments that such cultural artefacts 'brought the thoughts of other minds together: they externalised memory, enabled people to acquire far more from their social surroundings than they could ever hope to learn for themselves'.[15] These pre-ancestors made fire from flint stone, shaped clay figures of animals and humans, they made the first musical instruments and probably cried with grief. Burying the dead, with some afterlife in mind, implied a ceremony, which itself implied an emotional aura, and the recognition of loss, the beginnings of grief. MacLean even suggests a possible origin of emotional crying linked with cremation and the subsequent irritation of the eyes—a combination of sadness with smoky tears.[16]

THE SPLIT

Chapter 1 introduced the dichotomy between the two iconic gods Apollo and Dionysus, whose images were in Nietzsche's ideas grafted onto the psyche of the early Greek mind. However, similar characterizations were recorded in some cultures well before the golden age of Greece. Sometime in the 4th millennium BC, writing and mathematical measurement began to develop within human cultures, and with them came ideas of civil and cosmic order. Allegorical identifications came to be taken seriously as, probably for the first time, mythology became imbricated with reality and became codified within institutional agendas. Systematization and categorization became the overriding cultural progression, which took advantage of and enslaved the drives and cognitive structures of the human psyche.

The philosophical identification of dualism is often that of a distinction between two worlds, one transcendent, ultimately unknowable, and one knowable to and through the senses. These come down to us as variant dichotomies, such as those between spirit and matter or between sacred and profane. They have become embedded in the underlying myths that try to explain the origins and course of the universe. To quote Campbell: 'Thus a complete new mythology arose, and instead of the ancient Sumero-Babylonian contemplation of the disappearances and reappearances of planets as revelatory of an order of nature with which society was to be held in accord, an idea of good and evil, light and dark, even of life and death as separable took hold'[17]—darkness and light, the founding metaphors of Western philosophy.[18]

The sociologist Émile Durkheim (1858–1917), who viewed religion primarily from a sociological perspective, believed that by studying religion it was possible to get to the origins of human thought. Religious beliefs were only a special case of a very general law, namely that the whole of our social world is populated with forces that exist only in our minds. The mental life of mankind is a system of representations; the most commonplace objects can become sacred, and a totem becomes real. Feelings expressed collectively in a religious ceremony become dispersed soon after the event, but can be kept alive for an individual by being inscribed in symbolic form on things durable, such as totems and talismans, described in books, and written as poetry or composed as music. We all come to treasure our own special objects, mementoes of our past, which are evocative of people and events that have special significance for us, and which can easily provoke tears when touched or viewed.

Edward Wilson has written widely on the biological underpinnings of culture, embracing a holistic sociobiology. He considers that the emergence of civilization related to hypertrophy of pre-existing social structures, basic social functions of our ancestors metamorphosing via genetic advantage and evolutionary changes from environmental adaptations to elaborate social behaviours.[19] Drawing on anthropological data, Pascal Boyer suggested that attribution of agency is a common feature of the way the human mind works as an example of more general cognitive predispositions. We hear a noise outside on a dark night, and attribute it first to a predator rather than a branch of a tree falling. We constantly use such intuitions when interacting with others, and the attribution of minds to others is a crucial feature of human cognition. This has become referred to as the Theory of Mind, which is also referred to as mentalizing.[20]

ALTRUISM AND EMPATHY AND THE SOCIAL BRAIN

Richard Dawkins considers the evolutionary development of urges to kindness, such as altruism, kinship, and reciprocation—the giving in exchange for taking, along with generosity, empathy, and pity, as programmed into our brains.[21]

The ability to imagine that another has a mind the same as or similar to one's own develops at around the age of 4. With this arises the possibility that the mind of others differs from one's own, and also the possibility of trying to influence the other's mind and hence their beliefs. The underlying attribute may not be uniquely human, and may also be a part of chimpanzee cognition. However, combined with the power of human language, the development of a Theory of Mind is considered a significant development in the social evolution of *Homo sapiens*, since evolutionary selection must have favoured cerebral mechanisms that allowed for the rapid evaluation of the emotional state of a conspecific.[22]

The Theory of Mind is one component of what is referred to as 'the social brain', social cognition being defined as 'the ability to understand people's behaviour through the use of cues such as facial expression, eye gaze, body postures— including gesture— and social linguistic factors, such as prosody and the social content of speech'.[23] The ability to try to understand the mind of others allows for the development of a narrative, about what A believes about B, who considers what C believes about A, and so on. The ability to communicate and then write down such narratives leads to stories, of fiction, of history and myths, and allows for the projection of the contents of minds onto and into inanimate objects, and the evolution of art forms, including the stage play. To use Othello as an example given by Robin Dunbar, 'he had to *intend* that the audience *believed* that Iago *wanted* Othello to *suppose* that Desdemona *loved* Cassio and that Cassio, in turn, *loved* her (and hence that they planned to run away together)'.[24] Dunbar goes on to suggest that such storytelling helps bind us into groups, consolidate our sense of kinship, and promote the search for origins of the tribe.

Much research has been carried out on Theory of Mind, which skill seems to be absent in those suffering from autism,[25] and is dependent on a relatively discrete neuroanatomy, unravelled with brain imaging and well-designed psychological experiments. The neuropsychologist Chris Frith refers to Theory of Mind as a *cornerstone* of cultural evolution.[26]

But all this implies another mental attribute, namely that of mental time travel—the ability to think of and into the future.

To live in an imagined world, to coalesce the past with the future, to evoke the presence of absence, to fear the absence of presence—these may be truly human attributes too, although the extent to which any of these abilities are available to the mind of the apes is quite uncertain.[27]

The memory we have for everyday events is referred to as episodic memory. Some people consider that they have perfect memories, and believe that memory is laid down in the way a video recorder makes a recording of a television programme, detailing events carefully, which are not erased or changed with time. But this is as far removed from reality as reality is from our memories. As the writer Milan Kundera has expressed it, 'to remember is one way of forgetting,'[28] but it is also one way of creating. It is through the embodied traces of past events that we harmonize the immediate present with the past, refreshing the content anew each time we re-memorize, forming part of the ever developing narrative of our own personality. We image and plan for the future, but based on aoristic scripts. Sophisticated toolmaking in our hominid ancestors, going back for some 1.5 million years, is taken as evidence of future planning, but surely burial of the dead with artefacts at least 100,000 years ago is proof positive of both the presence of a Theory of Mind and planning for the future.

Episodic memories, tracing episodes, place the tragic fourth dimension, the awareness of time and its passing, into the primitive appreciation of the three-dimensional world.

MORE NEUROANATOMY

Mental time travel, reaching for the future as well as the past, emboldens all our lives and drives. It is of interest that recent memory research has shown that the brain structures that are essential for embedding episodic memories, especially the hippocampus, the surrounding cortex (parahippocampal gyrus), a part of the parietal cortex called the precuneus, and the right frontopolar prefrontal cortex, are also involved when travelling imaginatively in time.[29] One view is that the hippocampus stores information about where emotional events occur, the associated mood being given by amygdala activity, both being linked with the recall of emotional states in conjunction with time and place. This may involve the back projections of these limbic structures to wide regions of the cerebral cortex where representations of memories are stored. The fiction of the future is created from the faction of the past; remembering is re-membering—in several languages the verb is reflexive (French, *se rapeller*; German *sich erinnern*), remembering the self. In English this is evocative of members, of arms and legs, movements and actions, the embodiment of our memories and emotions.

Theory of Mind posits that an individual can imply the intentions of others, predict their behaviour, and recognize their emotions. Emotional contagion and perspective-taking are aspects that relate to empathy. Perspective taking is closely linked with Theory of Mind, cognitively and anatomically. Much research has been carried out on these attributes,

TABLE 2. **Regions of the brain involved in empathy, Theory of Mind, and mirror neurons**[1]

	Empathy	Theory of Mind	Mirror neurons
Medial prefrontal cortex	+[2]	+	X[3]
Broca's area[4]	+	X	+
Insula	+ (pain)	X	+
Amygdala	+	+	X
Anterior cingulate	+	+	X
Orbitofrontal cortex	X	+	X
Temporal pole[5]	?	+	X

Notes:

[1] This is not a complete list of brain areas implicated. For further information, see Carrington and Bailey, 'Are There Theory of Mind Regions in the Brain?'; Blakemore, 'Social Brain in Adolescence'; Blakemore, 'Developing Social Brain'. The area referred to as superior temporal sulcus is constantly reported as relevant, being linked to the perception of biological motion. For figures of the areas see Chapter 3.

[2] + = involved.

[3] X = not involved.

[4] Broca's area is part of the lateral frontal cortex which relates to language output, especially on the left side.

[5] The temporal pole is at the very front of the temporal lobe.

largely with new brain-imaging techniques, and the relevant overlapping brain areas for Theory of Mind and for empathy are shown in Table 2. The medial prefrontal and the orbitofrontal regions have been implicated in nearly all studies of Theory of Mind.

The importance of the orbitofrontal cortex for emotional expression has already been noted, but it is the medial prefrontal cortex which is especially relevant to the development of empathy and Theory of Mind in humans.[30] It is involved in such human feelings as guilt and embarrassment, and is activated by thinking about the psychological state of oneself or of others. The medial prefrontal cortex has increased hugely in size in humans compared to monkeys, although the size of the change is less clear in comparison to chimpanzees.[31] Unfortunately, there is little work to date on the development of Theory of Mind as children grow up. The ability to attribute agency to someone or something has been shown to develop at around eighteen months, infants by that age following an adults' gaze or looking towards a pointed direction.[32]

Simone Shamay-Tsoory has argued for two different cerebral circuits for empathy, based on imaging data and studies of patients with brain lesions of a variety of pathologies. Using an empathy scale (the Interpersonal Reactivity Index), a Theory of Mind test, and an emotional cognitive test equivalent to emotional empathy, she attempted to separate cognitive empathy—understanding someone else's point of view but not feeling with them—from emotional empathy, sharing feelings. Emotional empathy embraces emotional contagion, emotional recognition, and shared pain. Shamay-Tsoory thinks that, from an evolutionary point of view it develops prior to cognitive empathy; it is demonstrated earlier in young infants.

In her investigations, she found that emotional empathy involves the inferior frontal gyrus on the lateral surface of the brain and Broca's area, while cognitive empathy is associated with the medial prefrontal cortex.[33] Broca's area is part of the lateral frontal cortex, which is closely associated with spoken language, gesture, and imitation, while the medial prefrontal cortex is linked to Theory of Mind.[34] According to Richard Passingham, Broca's area is not unique to the human brain, and has been identified in monkeys and chimpanzees, while the frontal pole can be regarded as 'the top of the information processing hierarchy of the (human) brain', occupying roughly twice as much of total brain volume in the human as in the chimpanzee.[35] As noted, it is activated along with the hippocampus when subjects are asked to imagine future events.[36] Wildgruber and colleagues, using fMRI, reported that affective prosody, that is modulation of the tone of voice to convey emotion, elicited a right-hemisphere pattern of activation in areas homologous to Broca's area and frontal areas closely related to the frontal pole.[37] These anatomical associations link responses to emotional tone with empathy and the imagination of the future. The musicality of language, and the happiness and sadness of expression develop early in infancy, as does the ability to associate the minor key with sadness.[38]

These neuroanatomical circuits related to empathy can be shown to be activated, for example, when subjects are watching videos of people in pain, or watching a partner being subjected to physical pain, the activation correlating with individual empathy scores.[39] Shamay-Tsoory does not imply that in humans the two systems of empathy discussed by her are somehow independent, or that the neuronal responses

are unassociated with other structures related to the neurology of the emotions. Brain areas linked with these empathic
responses include the insula and anterior cingulate gyrus, elements of the circuitry involved in the anatomy of the emotions already discussed in Chapter 3. Part of the cingulate area
called the subgenual cingulate (area 25: see Figure 8 p62)[40]
receives strong projections from the amygdala, from the
insula, and from the pole of the temporal cortex. Since there
are direct outputs from the cingulate area to the solitary
nucleus and the dorsal motor nucleus of the vagus, the links
between empathy, emotion, and the generation of feelings
can be correlated neuroanatomically.[41]

It will be recalled how emotional crying is linked with
measures of empathy. The evolutionary development of cognitive empathy, related to Theory of Mind, with corresponding large increases in the size of areas of the human prefrontal
cortex, provides experimental and neuroanatomical evidence explaining, from a neurobiological perspective, the
human ability to *feel* the sadness of others, and to cry emotional tears.

MIRROR, MIRROR ON THE WALL

People cry when they observe others cry, as they laugh when
others laugh. This could be just mimicry, mere imitation, but
it is more often than not associated with the appropriate
feeling. As noted above, observing others in pain evokes
sensations in us that are referred to as empathic, and imply

compassion.[42] Empathic individuals are more inclined to mimic unconsciously the facial expressions of others,[43] and subjects imitating or just observing the emotional facial expressions of others show in ƒMRI studies enhanced activity in a network of structures that have to do with motor priming (preparation to act), but also in the insula and the amygdala.[44] In these studies, responses in the premotor regions were seen while observing emotional expressions, while those of the insula and the amygdala, notably on the right, were greater with imitation. These findings suggest that observing the emotion of another provokes action representation in the brain in areas to do with the face, and imitation of the emotion, via feedback from the facial muscles, enhances that emotional response. The latter suggestion is supported by observations that exposing people to pictures of emotional facial expressions for brief periods of time, for which they were not conscious, leads to activity, as measured with electromyography recordings, in distinct facial muscles that correspond to the appropriate emotion stimulus faces.[45] As Laurie Carr and her associates put it: 'to empathise, we need to invoke the representation of the actions associated with the emotions we are witnessing. In the human brain, this empathic resonance occurs via communication between action representation networks and limbic areas provided by the insula.'[46]

There have been several highly significant neuroscience discoveries in the post-Darwinian era that have gone beyond Darwin in realigning the position of *Homo sapiens* within the natural world of time and space. The neuroanatomical discoveries of the limbic structures and cerebral circuits which allow us to have emotional experiences are among those already discussed. Building on that neuroanatomy, the neuronal activities in areas such as the ventral striatum and amygdala have been shown to drive and bind our behaviours to social cues. Psychological constructs such as volition, motivation, addiction, craving, and what are referred to as reward reinforcers have now been revealed. Rolls notes that emotions may be elicited by primary reinforcers or by stimuli that become associated by learning with primary reinforcers called secondary reinforcers. As Rolls puts it: 'Many stimuli, such as the sight of an object, have no intrinsic emotional effect. They are not primary reinforcers. Yet they can come as a result of learning to have emotional significance.'[47] Facial expression is a primary reinforcer, as is parental attachment to infants and vice versa. Rolls also includes altruism in his list of primary reinforcers,[48] but it is through the secondary reinforcers that the bespoke gamut of the emotional behavioural repertoire becomes constructed, interlinked with environmental circumstances and social reinforcers.

There are other surprises. As the neuropsychologist Michael Gazzaniga puts it, 'The brain finishes the work half a second before the information it processes reaches our consciousness.'[49] Thus after a sensory stimulus, peripheral sensations reach the cortex of our brains rapidly, in some 20 ms, but they become conscious to us rather later, at a time lapse of around 500 ms. We then refer backwards in time, when considering the timing of the onset of the stimulus. Furthermore, Benjamin Libet and colleagues, in investigations which have been replicated several times, have revealed that our brain appears to be unconsciously making decisions for us to move even before we act, or become conscious of our intention to act. In these studies, subjects monitored with electroencephalography recordings were instructed to make volitional finger movements, and indicate when they actually decided to do so. An electrical potential, known as the readiness potential, appeared some 550 ms before the actual movement took place, but the moment of the individual's conscious intention was noted after another 350 ms. In other words, the brain's processing of the motor action occurs unconsciously, some 350 ms *before* the conscious awareness of the intent to action. Notwithstanding the controversies that such findings have caused, especially for arguments about free will and the nature of voluntary acts, they reveal that we live in the past, and even our treasured, thoughtfully self-willed intentions have unconscious beginnings.[50]

To these discoveries has to be added the emotional responses of the limbic-related structures, notably in the amygdala, which occur rapidly and without conscious awareness of the emotional stimulus that activated them. Using a technique of backward masking, in which a very briefly given emotionally laden image is masked by an immediate longer presentation of a neutral one, and the imaging technique of *f*MRI, Ray Dolan and his colleagues have shown that the amygdala responds to the former emotional but not consciously appreciated image, leading on to the neural cascade of the emotional system networks described in Chapter 3.[51] This pre-attentive emotional processing is just another example of the way the human brain processes our behaviours, leaving us consciously unaware of the whys and wherefores of our predispositions and actions, which we then have to justify by conformity to our own personal narratives. As the neuroscientist Marc Jeannerod puts it, 'consciousness...reads behaviour rather than starting it'.[52]

The latest assault on our human autonomy is the discovery in the 1990s of mirror neurons in the brain. The idea is an old one. Hume wrote: 'the minds of men are mirrors to one another, not only because they reflect each other's emotions, but also because those rays of passions, sentiments and opinions may be often reverberated and may decay away by insensible degrees.'[53] Nietzsche could almost be referring to the importance of mirror neurons when he comments on empathy in Tragedy: 'to begin to act as if one had actually entered into another body, another character. This stands at the beginning of development of drama...here we have a surrender of individuality.'[54]

In investigations using monkeys, Giacomo Rizzolatti and his group first observed that an area of premotor cortex could be activated not only when the animal made grasping movements, but also when it observed another animal making the same movements. This area (referred to as area F5 in the monkey brain) is considered to be a homologue of Broca's area in the human brain. In other words, the same sets of neurons are activated in the brain when an individual does something, as when they only see the same act performed by another. However, they respond only to actions that have biological 'meaning', in other words not to seeing the grasping action of a mechanical implement. From a biological perspective, the interpretation and meaning of these findings remain controversial, but 'mirror' neurons have been recorded in many other experimental situations, and are thought to be involved in the development of Theory of Mind.[55] Initially the observations related mainly to manual actions, but then mirror neurons were identified in respect of the auditory system, and with states of emotion. Further, their presence in the human brain has been well established. They have been observed in premotor areas in response to seeing emotions in others, as noted already, the responses being confined to behaviours of relevance for the individual. It has been noted that mimicry of an emotion can activate the same cortical areas as observing it, and it is relevant that the involved cerebral structures link with the mirror neuron system affecting the premotor areas, the superior temporal cortex, the insula, and the amygdala. Buccino and colleagues noted activations in humans in response to human oral gestures of speech, but not to images of a monkey lip-smacking or of a dog barking.[56] These neurons seem to be triggered to action very quickly, and thus reveal an unconscious system for monitoring the intentions of others.

There are no studies as yet on the activity of mirror neurons in relation to crying, but some studies have looked at other emotions. Viewing images of the expression of disgust and inhaling odours that provoke disgust have been shown to activate the same parts of the insula, the viscerosensory cortex at the vortex of our emotional feelings.[57]

The insula is consistently activated in studies of emotion, including social emotions, and it is of relevance that the insula receives information from that part of the temporal cortex which responds to the sight of faces. The underlying anatomy of facial recognition can be briefly outlined.[58] Facial recognition is dependent on visual input from the visual areas of the cortex projecting via several cascades of information of increasing complexity, to the inferior temporal cortex, which projects to the amygdala, the hippocampus, and thence to the ventral striatum (see Figures 4 p 55, and 5 p 56). Certain parts of the amygdala, known to be associated with facial recognition, have increased in size in primates in relation to the size of the social group.[59] Since the amygdala is closely associated with the expression of emotion, recognition (as re-cognition) and expression of emotion go hand in hand. Information regarding facial expression is also provided to the orbitofrontal cortex, shown to contain neurons that respond to primary and secondary reinforcers, outputs from which alter activity in the hypothalamus, the cingulate cortex, and the autonomic nuclei of the brainstem (see Figure 11 p 71).[60] These circuits are influenced by cognitive states, the latter altering emotional representations in the brain, enhancing the emotional state. According to Rolls, emotions are represented in the orbitofrontal cortex, and the activity of this cortex is influenced by language. Thus, as an example, he cites investigations of his group which showed how the interpretation of pleasantness of the smell of an ambiguous olfactory stimulus was affected by words delivered during the olfaction, and this effect correlated with associated fMRI findings in the orbitofrontal cortex. These studies show

> that cognitive influences, originating from as high in processing as linguistic representations, can reach down into the first part of the brain in which emotion, affective, hedonic or reward value is made explicit in the representation, the orbitofrontal cortex, to modulate the responses there to affective stimuli...linguistic representations can influence how emotional states are represented and thus experienced.[61]

Viewing an emotion thus activates the neuronal core of the individual's own experience of the emotion, as predicted by the theories of Damasio. But there is more, since the orbitofrontal cortex has neurones that are responsive to musical sounds.[62] With music we resound (re-sound) with feelings and resonate with empathy, with emotional contagion, in settings of tragedy and Tragedy.

With regard to the lateralization of these responses to facial recognition, the relative contributions of the two hemispheres are unclear, but most investigators favour the right hemisphere as the one most involved in the processing of Theory of Mind, emotional prosody, emotional words that reference the self, recall of personal episodic memories, and prospective mental time travel.[63]

The discovery of mirror neurons, along with other findings such as the discovery of the readiness potential described above, have altered the way that neuroscientists view the function of motor neurons, or even the motor system generally. The traditional view is that perception precedes motor action: seeing something triggers the motor response. However, there is growing evidence that there is a motor pre-presentation preceding the sensory input, which is a part of the perceptual filtering process, a readiness or protension that guides not only what it is that we perceive, but even what it is that we want to perceive. Marc Jeannerod refers to this as 'motor cognition'.[64] In effect the brain anticipates the immediate surroundings: it is 'projacent', as the philosopher John Searle termed it, and we use intentions in action, not reactions.[65] Another philosopher, Maurice Merleau-Ponty (1908–1961) used the term 'motor intentionality' to describe the way in which the body directs itself towards, and grasps, objects in a precognitive manner. Motor activities occur in environments, and the actions we make in most circumstances are purposeful—we perceive opportunities to carry out acts; the acts 'feel' familiar; we *know* how to touch the world around us; our bodies coincide with the world as we have coition with things and with others. In Merleau-Ponty's philosophy, perception is a *motor skill*, a fundamental change of perspective for philosophy, psychology, and neurology.[66]

After an action, feedback from our muscles monitors information about the state of the action, adjusting the next part of the execution to the intention. This is a 'forward-seeking model', but according to Jeannerod it applies not only to activation of the voluntary muscles but also to the autonomic nervous system.[67] It has been shown, for example, that during tasks of imagining the motor acts related to graded changes of effort, such as peddling a bicycle at various rates, autonomic changes such as increased heart and respiration rates are recorded prior to any increase in muscle metabolism. Jeannerod refers to this as an uncontrolled 'leaking' of the central processes of action representation.[68] This is unconscious and Jeannerod was further able to show that activation of the autonomic nervous system is far harder to inhibit than activation of a voluntary motor pathway during motor imagery.

These motor cognitions reflect embodied mental states, which can be triggered by viewing the actions of others. As Vitorio Gallese and Alvir Goldman reflect: 'when one is observing the actions of another, one undergoes a neural event that is qualitatively the same as (the) event that triggers actual movement in the observed agent...a mind-reader represents an actor's behaviour by representing in himself the plans or movement intentions of the actor.'[69] As the neuropsychiatrist James Harris puts it, 'our empathic resonance is grounded in the experience of our bodies in action and the emotions associated with specific bodily movements'.[70] We live in three-dimensional space, and relate to it with our bodies;

there is no disembodied Cartesian empire which occupies some space which is no place. These neuronal events relating to emotion occur unconsciously, and are both automatic and pre-reflexive: to see another crying affects us. As the Roman poet Horace put it so neatly:

> If thou wish me to weep,
> Thou must first shed tears thyself;
> Then thy sorrows will touch me.[71]

FINALLY, THE FAMILY

Maclean's point with regard to his statement that the history of the evolution of the limbic system is the history of the evolution of the mammals, and hence the family, was that with the evolution of mammals came infant–mother attachment and bonding, which became more complex as the phylogeny advanced through to *Homo sapiens*, possessing the most elaborate limbic system, and most complex social structures.

It can be argued that the human brain expanded alongside an ability to cope with the complexity of living in larger social groups, with the necessary development of empathy and Theory of Mind for social coherence. Also, during primate evolution, the shift in environment from tree tops to the savannah, from living in groups to becoming hunter-gatherers, enforced the importance of greater visual and auditory detection of emotions. This was enhanced as the muscles of facial expression and the cerebral anatomy relating to them developed and the social engagement system enhanced the detection of the emotions of others.

Harris among others has emphasized the importance of empathy in the evolution of mankind, uniting MacLean's ideas with those of Porges. Porges relates the emergence of empathy with the increasing complexity of the autonomic nervous system, with the sophistication of the social engagement system, and with a shift to audio-vocal communication with the evolution of the mammalian middle ear.[72] He makes the point that the human middle ear developed from the jaw bones of earlier reptiles, the detached middle ear bone being one of the defining anatomical features of mammals. The human middle ear carries sound at only specific frequencies. It is naturally attuned to the sound of the human voice, and the frequency band corresponds to that which composers have traditionally used to compose melodies, and which mothers use to sing to their babies.

Corballis speculates that at some point in primate evolution, facial expressions overtook hand gesture for significant communication, freeing the hands for manipulation and manufacture.[73] Ridley noted the importance of artefacts and thence of art itself for cultural and social advancement.[74] With the addition of increased powers of vocalization, language as we know it developed, which could be combined with the capability of neurones to mirror the actions of others, the attribution of agency by the mind, as well as the ability to mentally travel in time to a future. When our ancestors moved to living in the open landscape, safety in numbers promoted the adhesion of larger social groups, and with meat eating made easier and more enjoyable by fire and hence cooking, social communication, especially around the fireplace, must have increased. This was perhaps the beginning of sharing, of times and places, of past and possible future events, and of ritual, all occurring perhaps somewhere along a time axis from two million to 200,000 years ago.[75]

Rolls refers to social bonding as a primary, gene-specified reinforcer of behaviour,[76] and altruism is often quoted as an example of behaviour that leads to species selection, where what is best for the individual coincides with that which is best for the survival of the group. Ridley argues for a genetic change of brain structure occurring some 200,000–300,000 years ago, a difference given by small changes in gene composition, but subtly and importantly altering gene expression and related to significant changes of social and cultural behaviours.[77]

Gana, the heartbroken gorilla, cradled her dead baby but did not cry. Apparently, female apes cling to their babies and rarely put them down, but in humans, direct communication between a mother and infant can continue without clinging together, through vocal and facial communication, which allows mothers freedom from carrying their child early on after birth. There is also the importance of neotony. This is the biological and evolutionary process which has led to the prolonged period of dependency of the infant and growing child on a carer or carers before achieving independence. Our development takes much longer than any other known species, and a lot of that time is given to engaging with conspecifics, learning to live and love. The importance of the separation cry in the course of mammalian development has been discussed. Emotional crying too serves as a primary reinforcer in the infant–parent setting, provoking an emotional response in both the parent and the infant. The former is one of 'attachment', while the latter is intimately bound with this. When the appropriate action is taken by the parent, comfort, relief of stress, and the cessation of tears follow.

Lucy, perhaps our oldest known Australopithecine ancestor, wandering on two legs and scavenging in the African savannah some three million years ago, left us the legacy of her bones, but we do not know for how long she held her infant, or if she cried tears.[78] The epoch of *Homo ergaster*, some 1.5 million years ago, saw a reduction in male size in comparison to females, and lasting monogamous bonds between men and women are thought to have evolved then.[79] The latter must have involved the development of a more complex Theory of Mind than had previously existed, but also increasing empathy and an ability to better envision the future. The loss of body hair in *Homo sapiens* would also have helped, as body adornment and increased flexibility of facial expressions became more apparent. In combination with language, these changes would have helped usher in an astonishing richness of culture far beyond that of earlier hominids.

5

TRAGEDY AND TEARS

...in a gushing stream
The tears rushed forth from her o'erclouded brain,
Like mountain mists at length dissolved in rain.

Lord Byron, *Don Juan*[1]

RECAPITULATION

The views expressed by Nietzsche in *The Birth of Tragedy* contrast with the usual purely literary or sociological theories. Tragedy, in the traditional view, is seen as a special literary form, which has been given certain definable characteristics. Arising in Greece in the 5th century BC, it referred to plays written for festivals, most notably by tragedians such as Aeschylus, Sophocles, and Euripides. There may have been good reasons why Tragedy became such a distinctive art form at the time it did, in relation to the wealth and the political and social structures of Athens in the 5th century BC. However, if we are to understand how Homer's heroes hovered on the slopes of the Parthenon,[2] we need to reach back to an earlier era, when more elementary social structures regulated communities; when gods and myths were all-pervasive in the quotidian life and still developing social cognition of *Homo sapiens*.

Unfortunately we have little documentary evidence of the cultural displays and religious ceremonies that marked the cycles of life in such early times, or what kind of imagination may have created them. But we can be sure that they were created, enacted, and gradually evolved, with the developing sophistication of the growing size of communities and the rise of the earliest cities.

The first writer to try to define the characteristics of Tragedy was the philosopher Aristotle (384–322 BC).[3] His views, based largely on the plays of Sophocles, have influenced discussion of the subject ever since. Many would say that Aristotle's strict interpretations of Tragedy have hindered rather than helped our understanding of the genre, and they have been challenged by many, including Nietzsche. *The Birth of Tragedy* itself has also attracted coruscating criticism, but it inspired a renewed interest in examining the origins of Tragedy, though from a rather different perspective.

One of the scandals of Nietzsche's interpretation was his undermining of the classical ideals of Greek Tragedy from its pedestal and his placing of archaic rituals in the driving seat. The latter were closely bound up with control over the cycles of nature, which involved animal or even human sacrifice. They linked with concepts of dismemberment, death, and resurrection with the annual cycles of the seasons, and were played out in rituals and ceremonies. As noted by scholars such as Campbell and James Frazer (1854–1941),[4] such events, whether in honour of Osiris, Adonis, Orpheus, or Dionysus, are found universally in evolutionary historicism.[5] They pre-dated the era of classical Tragedy, and even if the term 'the-atre' poorly reflects their enactments, the content, as far as we know, was communal and religious, and portrayed life and its continuance within the context of death.

These views imply a ritualistic origin for drama, but while ritual is aimed at the gods, drama is for humans. The idea that such a pure aesthetic entity as Tragedy could have evolved out of primitive rituals was not approved of by those who revered the Greeks and their high culture as something entirely new in the intellectual history of mankind. Their thinking lionized that era and the early classical philosophers. This view expressed a need to return to that golden age, a prelapsarian Eden, from which all successors had fallen.[6] However, dramas with a narrative plot involving key individuals (ur-heroes) and human suffering developed at times and places well before Greece in the 5th century BC. Aristotle called drama, literally 'something done'—a passive noun—a praxis, a doing, an action; he saw it as an active event. In Athens at the time Tragedy became a civic event, with social and philosophical importance, the gods, although present, took more of a back seat. In the extant classical tragedies, Dionysus has almost vanished from the stage except in *The Bacchae* of Euripides. But for Nietzsche, Dionysus was the energy in the Tragedy; in *The Bacchae* it is he who sets Thebes dancing.[7] Tragedy surely emerged from US-rituals, archetypes related to the development of the evolving human psyche.

TRAGEDY BRINGS FORTH TEARS

Given the volumes of books and papers written on the emotions, the dearth of interest in crying or the emotions in Tragedy, except from a purely literary point of view, is remarkable.[8] Research on the where and when of emotional crying rehearses certain recurrent themes, discussed in Chapter 2. These include times of mourning, listening to music, singing, institutional ceremony, and religious occasions. With regards to the arts, crying as a response, which is so closely tied to music, takes place everywhere, in the home, the theatre, the cinema, and other settings. Theories about the function of crying have swayed from the biological—getting rid of bad humours—to the psychoanalytic; and some pivot around the word 'catharsis' and the ideas of Aristotle. If anything, the physiological responses to crying as reviewed in Chapter 2 reveal *increased* autonomic activity, rather than any obvious purging, an arousal rather than a calming effect.

We have now looked at the communicative value of crying, especially within the evolutionary psychobiology of writers such as MacLean, Linking it with social bonding, sympathy, and empathy, and the underlying neuroanatomy of crying revealed by contemporary neuroscience has been described. While this book is about tears shed emotionally, and while it is obviously possible to approach this from purely sociological or psychological perspectives, a full understanding of human behaviour, from the simplest to the most complex, must embrace both evolutionary perspectives and neurobiology.

NEUROANATOMY—AGAIN

The most important concepts for an understanding of the neuroanatomy of our emotions are as follows. First, the brain of *Homo sapiens* is the outcome of over a billion years of evolution: change and metamorphosis eventually forged the eloquent neocortex of humankind. The present form should certainly not be considered final. Any philosophy that considers twenty-first-century *Homo sapiens* to be the endpoint of some teleological progression fails to recognize the nature of Darwinian evolution. Secondly, recent neuroanatomical findings reveal close links between those parts of the brain referred to as limbic (very, very old), and the neocortex, so called because it is newer, but in reality older than is usually understood. The neocortex is very much imbricated with and driven by limbic neural activity—emotion underpinning motion and mind, and forging cognition. The dominance of the old, limbic, and subcortical structures over neocortical activity in regulating behaviour ensures a continual triumph of the emotional over the rational. This seems self-evident if we place the human brain in the context of millions of years of evolutionary adaptation, during which time individual survival depended on rapid evaluation of the environment and swift action based on emotional appraisal of any given situation. Fight or flight, no time to think! In contemporary neuroscience, emotion is no longer seen as a counterpart to reason in human cognition, but as a collaborator and indeed coeval constructor of our reasoning and thinking.

Thirdly, the importance of the autonomic system and its complicated underlying anatomy is central. Two essential findings discussed in Chapters 3 and 4 were the development of the social engagement system and neuroception in *Homo sapiens*, along with the corresponding reorganization and expansion of the autonomic control of facial and vocal expression. The reorganization and coordination of the pharynx, larynx, and thus breathing and heart rate were essential for smiling, laughing, and emotional crying, as was the *direct* influence of the neocortex on the autonomic nuclei in the brainstem. The key cortical structures involved are the orbital and medial frontal cortex, the insula, and the anterior cingulate cortex, with an emphasis on the right hemisphere of the brain.

Fourthly, the neuroanatomy underpinning empathy, Theory of Mind and crying (discussed in Chapter 4) takes the neurobiology linked with emotional crying a step further. This step may seem small in terms of the alterations in brain size involved, but it represents one huge step for mankind— the neurobiological basis of a morality and ethics which has concerned all known civilizations. It is clear that apes, and perhaps some other primates, show preliminary evidence of altruistic behaviour. An often cited case of what is referred to as reciprocal altruism is that of capuchin monkeys, who can be shown to share more with those who have helped them with grooming than others, and appear slighted if they see another receiving a better reward than they themselves receive for performing the same task.[9] There are several authors who cite examples of what they consider to be altruistic behaviour in species as far removed from each other as mice and grizzly bears.[10]

Fifthly, the possible origins of languages have been examined. No one knows how, when, and where languages as we know them today arose, although speculations abound. The view taken here is that this remarkable event emerged via a proto-language, driven by gesture, framed by musicality, and performed with the flexibility which has accrued with expanded anatomical developments, not only in our cerebral cortex, but also our facial, pharyngeal, and laryngeal muscles. Around the same time (with a precision of many thousands of years), the bicameral brain, while remaining bipartite, with the two cooperating cerebral hemispheres coordinating life for the individual in cohesion with the environment, became differently balanced with regards to the functions of the two sides, the left and right: pointing and proposition (left), as opposed to urging and yearning (right).

From this came the formation and use of metaphors and then concepts, the instigation of boundaries and borders, of rules and regulations, of proscriptions and prohibitions, of writing and textual authority, of law and religion. Self-consciousness, awareness of other minds, bonding between individuals, feelings of empathy and love, and knowledge of death of the other and of the self descended upon the happy— some would say hapless—hominid. As Achilles said at the sight of the broken king Priam, after the death of his son Hector: 'for the gods have spun a thread for pitiful humanity, that the life of man should be sorrow'.[11]

The neuroanatomy underpinning these human cultural events is much clearer now than ever before. The limbic system captures the entire stream of sensory experience, and, unlike the neocortex, projects extensively to the hypothalamus and the pontine and mid-brain nuclei that regulate autonomic, endocrine, and motor output. The insula—the visceral sensory cortex capturing gut feelings—is central to this circuitry, also sending information about these feelings to the hypothalamus and beyond. Some neocortical areas such as the ventromedial prefrontal cortex and frontal pole have been associated with empathy, Theory of Mind, and mental time travel, while the equivalent of Broca's area, especially on the right side— the homologue of the main propositional language area on the left—is part of a mirror neuron system which is activated by seeing emotions. The non-dominant frontal cortex, aided by the hippocampus, adds context and memories, an amalgam of the past, the present, and the future.

From a neuroanatomical perspective there are overlaps in the brain circuits that are related to four essential components of our understanding of the emotional responses to tragedy, and hence Tragedy. These are Theory of Mind, mirror neurons, empathy, and mental time travel. The neuroanatomy linking these are the ventromedial prefrontal cortex (mental time travel, empathy, and Theory of Mind), the insula (gut feelings, mirror neurons), and the Broca's area homologue (mirror neurons and empathy).[12]

Brain-imaging investigations reveal a substantial role for the right hemisphere in modulating some very important aspects of language and also memory. The memory system involved is personal, and the linguistic skills are related to metaphor and idiom interpretation, especially the ability to recognize alternative meanings of phrases, and those statements that defy the literal rules of the language. The right hemisphere has been shown to have a dominant role in the processing of emotional language. One author, Norman Cook, summarized these achievements as follows:

> At every level of linguistic processing that has been investigated experimentally, the right hemisphere has been found to make characteristic contributions, from the processing of affective aspects of intonation, through the appreciation of word connotations, the decoding of the meaning of metaphors and figures of speech, to the understanding of the overall coherency of verbal humour, paragraphs and short stories.[13]

The important role of the right hemisphere of the brain in modulating, managing, and motivating human behaviour has been explored in detail only recently, and has had few champions in comparison with the propositional, indicative, language-rich left hemisphere. In the 1970s, Julian Jaynes (1920–97), in one of the earliest books from a new generation of writers looking at the brain and conscious experience, developed a theory that associated the right hemisphere of the bicameral brain in our ancestors with auditory hallucinations, literally the voices of the gods, a state which, he suggested, existed until the bicameral brain broke down around the time of Homeric writing. The heroes of the *Iliad* portray action, not willed self-consciousness; they are driven by the instructions of the gods, not by their individual personae. An evolutionary push then gave the left hemisphere access to the right hemisphere's voices; the latter became the object of subjectivity; written codes and narrative came to dominate; and the verbal abilities of the individual ego obscured the intuitive actions of the right hemisphere.[14]

Jaynes's views do not stand up to what we now know about cerebral anatomy and development, nor for that matter as a fine critique of early Greek epics, but one of his contributions was to place the study of human consciousness and aesthetics firmly in the context of the brain, notably drawing attention to the right hemisphere. This endeavour has since been taken up by others, as already discussed.[15] McGilchrist traces the progressive dulling of the right hemisphere's activities by those of the dominant left hemisphere, with the tragic consequences this has had for the Western mind and its associated culture. In *The Master and his Emissary* he reviews much neuroscience literature on the importance of right hemisphere activity for attention to things around us, to presenting a global, holistic contextual view of the lived-in world, one that is not re-represented, as is the world-view of the left hemisphere. The human right hemisphere is more concerned with living things and the capacity for empathy: 'it prioritizes whatever actually *is* and what concerns us. It prefers existing things, real scenes, and stimuli that can be made sense of in terms of the lived world, whatever it is that has meaning and value for us as human beings.'[16] Having more abundant connections to limbic structures, the right hemisphere is more in tune with the body's emotional physiology; it is intimately bound to the self, and bound to others through shared emotional empathy.[17] The neurologist Orrin Devinsky refers to the right hemisphere's sense of narrative, of existing in time, as the 'glue holding together the sense of self'.[18] McGilchrist emphasizes the primacy of wholeness, the right hemisphere dealing with the world before separation, division, and analysis has transformed it into something else—the Dionysian before the Apollonian.

TRAGIC JOY

Aristotle's concept of catharsis does not stand up to close scrutiny as the defining emotional effect of Tragedy. The effect of Tragedy is not to send people home sober and becalmed, nor, to take on a later meaning, is it to create a feeling of being purged from sin by seeing suffering and redemption in the protagonist. The response is often one of tears, and the feelings are more than just satisfaction. There is no satisfactory word to encompass them.

Over time, various authors have tried to parcel out the emotions into different categories. As with the histological divisions of the cerebral cortex, the number of different emotions described has varied. For James there were four, for Kant five, for Descartes and Darwin six, and the list is greater in the schemes of Rolls and Damasio. Some take a categorical approach, others a dimensional one. While all this may seem laudable from an empirical scientific perspective, it fails to capture all the subtle nuances of human emotional reflections. As Susanne Langer put it,

> verbal statement, which is our normal and most reliable means of communication, is almost useless for conveying knowledge about the precise character of affective life...any exact concepts of feeling and emotion cannot be projected into the logical form of literary language...the myriad forms of subjectivity, the infinitely complex sense of life, cannot be rendered linguistically.[19]

While all who pronounce on the schemes cover sadness or sorrow, none seem to include love, crying, the special affect associated with tears, or the feelings associated with Tragedy.

Many have discussed the nature of this feeling, 'the tragic emotion' as James Joyce referred to it. Nietzsche used the term 'tragic pathos'. Most discussions reverberate back to Aristotle who considered that to be effective, Tragedy should provoke pity and fear, which lead to purgation of the emotions and hence pleasure. There has been much debate, not only about the meaning of the two words 'pity' and 'fear' as used by Aristotle, but also about the translations of the original Greek ones (*eleos* and *phobos* respectively). In the extant writings, *eleos* is found in close association with the Greek words for tears and crying.[20]

Pity and fear—but who is pitied, and what is feared? While pity readily brings on tears, fear usually does not. How can a member of the audience pity Oedipus and fear marrying his own mother? Are we purged or purified? The philosopher Jonathan Barnes argues that the effects of Tragedy on him (and his friends) are not those identified by Aristotle, but that 'we are always at the mercy of chance, and it is precisely this aspect of the human situation which is the stuff of tragedy'.[21] For George Steiner there is 'a fusion of grief and joy, of lament over the fall of man and of rejoicing in the resurrection of his spirit.' This led to 'the momentary shock, the shiver in the spine—what the Romantics called *le frisson*—not the abiding terror of tragedy'.[22] Others refer to Tragedy inducing a sense of the sublime, or of aesthetic pain, with a combined emotion of debasement and at the same time awe, elevation, or grandeur. Roger Scruton conceives of Tragic theatre as recreating the experience of the liturgy, a 'recreation of this sacramental moment', a communion, but one which evokes 'primeval feelings of guilt, threat, and collective vengeance, and also the transition from sacrificial victim to sacred presence which is the gift of so many religions'. The unity between tragic and religious feelings is 'the experience of the sacred (being) a human universal, bound up with our very existence as self-conscious, rationally choosing subjects'.[23]

I suggest that the tragic feeling does not arise from a combination of known feelings, but from a different emotion, one that intimately concerns the self, infused with the vibrant remnants of our evolutionary past. This is sometimes called tragic qualm or tragic grief. Let us here call it *tragic joy*. To observe a calamity in real life is upsetting and may evoke pity, or sorrow, empathically felt. To be placed in personal danger may lead to feelings of anxiety and fear. However, the evoked emotion from witnessed Tragedy is different. I suggest here that it is essentially an ancient affect, generated from the tensions in the play and resolved by the closure of the plot, but with evolutionary origins. As Brian Boyd lays out in his *On the Origin of Stories*, telling tales is not only a human universal, but possibly even linked to a mammalian mode of understanding, arising with the evolution of memory, intelligence, language, and Theory of Mind.[24]

Of all the human passions, fear, anger, sorrow, and bereavement stand out as among the universal. They are deeply entwined into the stories of many Tragedies, but are also bound together in human tragedy. Fear and anger are emotions heavily dependent on limbic structures, especially the amygdala, the limbic structure that is central to our appraisal and expression of emotion, and, as will be revealed, which is perhaps a key to understanding this special feeling of tragic joy and the tears that are so often a part of it. Fear has an evolutionary history far older than mankind: all mammals know fear, and seem to express anger under appropriate circumstances. But grief and mourning have a cognitive component that is quite different, and it seems that, to use a phrase of Robert Burns', 'man was made to mourn'.[25]

Grieving is about loss, and it is often about the loss of a close one, although it may be for lost time, lost place, or lost opportunity. Where loss involves a person or persons, empathy is involved, as is an acute awareness of the future, emptiness, and the potential of others to mourn for you. In preliterate societies, the development of empathy and the ability to appreciate that other individuals have minds (Theory of Mind) would seem to have been crucial to higher hominid social cohesion, and hence to the evolution of *Homo sapiens*. Through the medium of early religious ceremonies and rights, fear became shared, and fear of death became mollified with the potential of the other world and an afterlife; mourning was part of this process (*Homo religiosus*). As a new type of consciousness emerged with the development of human language, and the words 'I', 'now,' and 'here' reified the individual's location in time and space, artefacts representative of our cognitions were created. This new form of creativity blossomed into art, initially of a religious nature, which enabled a binding of the individual in the present to the past and the future.[26]

BACK TO APOLLO AND DIONYSUS

So what can be said of the appellations Apollo and Dionysus which takes us away from simplistic dualisms? The idea of the two Greek gods as emblems of artistic intent was brought to its apotheosis by Nietzsche, for whom Tragedy arose from the irrational not the rational. As Steiner put it, 'Tragic drama tells us that the spheres of reason, order and justice are terribly limited'.[27] Portrayed as two gods in classical statues, Apollo and Dionysus have interest and beauty for us, but they cannot directly resonate from the past to the present. As metaphors for contrasts between the visual (painting and sculpture) and auditory (poetry and music), between the rational and irrational, or between the straight line and the curve, they can enliven our imagination. As myths for our time, they may serve in a similar way to how they aided the Greeks. We, like the Greeks, need myth and artistic symbols to convey something that is otherwise difficult to express in words.

The rise of Romanticism reflected a confirmation not only of the forces that underpin human activity in the cosmos, but also of the need of the individual human spirit to cope with the contingencies of existence. The dominance of unconscious drives within the human psyche was extensively discussed by Nietzsche, but revealed in full force by Freud, who suggested a kind of archaic heritage driving human behaviour, and by Jung, whose descriptions of archetypes reflect innate universal forms of behaviour which were residues of primitive evolutionary psychology. Indeed, the infusion of the unconscious into theories of Tragedy in the post-Freudian era was presented by Anthony Nuttall. For him the tragic effect reflects 'a quasi-physiological cathexis of psychic force'—catharsis to cathexis: the passive to the active.[28]

Apollo and Dionysus can be seen as aspects of the human psyche, representative of cognitive archetypes embedded in our neural past. Dionysus may have been a latecomer to the Olympian theocracy, but the energy and forces that he represents have been in the driving seat of the evolution of the hominid mind for millennia. As such, the personae of both gods are identifiable to each one of us. But they may also be seen as reflecting psychological attributes present within the developing psyche of our hominid ancestors, driving them to artistic creativity which was later captured in archaic religious rituals and then cultivated as the art of Tragedy.

Life for the Greeks was presumably little different in fundamental respects from what it is for us today. The world around them was unpredictable, cruel, and alien, one from which only optimism could provide an escape: optimism as reflected in a belief in inevitable progress, in a better existence to come, and in some form of salvation or redemption from those (Dionysian) forces. Art was for Nietzsche one such escape; for Socrates the embrace of the logos was another. Oedipus, discovering the truth, and trapped in his own destiny, struck out his own eyes. Unable to cry tears, his eyes wept blood.[29]

6

TEARFUL LOGIC

A curse first on the high pretences
Of our own intellectual pride!
A curse on our deluded senses
That keep life's surface beautified.

Goethe, Faust[1]

Prometheus, in Aeschylus' play *Prometheus Bound*, brought fire to humankind, but also taught them the arts and sciences. The German word *Kunst*, which is related to the verb *können* (to be able to do something), refers to both technical skill and art, hinting at the link between the two. *Kunsttrieb*, or creative drive, can be seen as representing a physiological force of nature, necessary for survival, honing hominid cognitions with skills to control the environment, which at a point in evolutionary development breaks out as symbolic representation and art. These ideas make the body's physiological processes fundamental, not only for the rise of consciousness but also for knowledge, reasoning, and creativity. Mainstream Western philosophy, as summed up by George Lakoff and Mark Johnson, has long clung to a view that rational thought is conscious, logical, transcendent, and dispassionate. Individual minds and the objective world were viewed as organized rationally, reason was a universal attribute, the enlightenment of the Enlightenment. The world was one in which things described (essences) endured, including the unity of consciousness, and the self was conceptualized as an unchanging entity, the subject as object. As Lakoff and Johnson argue, since abstract thoughts are largely metaphorical, and since so much metaphor is shaped by our bodily interactions with the world, our lived experience, rationality itself, is embodied.[2] Many metaphorical expressions come from physical encounters with the environment, for example, feeling 'up' or 'down' as metaphors for our body posture associated with such emotions. Rational thought is not disembodied, found in some ether-floating Cartesian ego. Concept formation is embedded through the body during ontogeny, and, as Damasio's theories imply, emotions are integral to it. As Damasio comments, 'our brains receive signals from deep in the living flesh and thus provide local as well as global maps of the intimate anatomy and intimate functional state of the living flesh'.[3]

Here are the views of the anthropologist Clive Finlayson:

> the brain developed from a geographical mapping organ, to one capable of much more. There was a by-product: it was

awareness of ourselves, not something unique to us…self-awareness comparable to that in humans now seems confirmed in elephants, bottlenose dolphins, and apes. Awareness of self would seem to be a natural consequence of awareness of objects in space and time, including other members of one's own species…it produced an animal capable of situating itself in space and time, an animal that became aware of the consequences of its own behaviour and mortality.[4]

MORE ON ANIMAL LIFE

To think that animals, especially primates, have no concepts about the world they live in, simply because they have no words, is strange, yet few philosophers consider the evolutionary backdrop to cognition in their works. During evolutionary development, the brain has adapted sensory receptors to identify boundaries, and hence objects, to give them permanence and to distinguish objects that have important meaning for the individual from those with less or no meaning. Categorization, the ability of animals to dissect out things in their environment, to distinguish one object or sensation from another, is a fundamental biological attribute of the first living creatures (to eat, or not to eat, or to be eaten). Knowledge of the world and how to manipulate or even classify objects in the world is a feature available to many animals, and object permanence has been clearly demonstrated in higher primates—namely the ability to track the position of an object removed from sight. (For instance, an object is placed in a box, the box is then taken behind a screen, where the object is removed from the box, and the empty box is then shown to the subject. The test is whether the subject then seeks the object behind the screen.) Object identification and utilization, and biological significance, are features of the developed mammal brain.

The brain is a pattern detector, and identifying patterns in nature allows for the quick assimilation of data for a nervous system primed to receive them. The patterns pursued are those that aid survival, that are reflected from and reflect what, at any particular time, is being sought; but the images are not passively received—they are actively grasped.[5] An organism's premotor predispositions are basic to the development of concepts. Damasio supports such views, arguing that an embodied concept is a neural structure embedded in the sensorimotor systems of the brain, such that conceptual inference is sensorimotor inference—we are linked to the world about us through our embodied interactions.[6] But breaking up auditory time and visual space to crack open meaning is a precursor of the Apollonian in the mind of *Homo sapiens*. Thus to deny that animals, especially our closest evolutionary relatives, have an 'inner world' of constructs, and emotional drives that are embedded and embodied in their cognitions, would seem obscure. As Hurford commented: 'Rudimentary concepts, ideas, and thoughts (or something like them), about things, events, and situations in the world, can reasonably be said to exist in animals' minds…The aboutness or Intentionality, of modern human utterances derives from the aboutness or Intentionality of pre-linguistic mental representations.[7] Boundaries, lines, objects—these are all identified within an animal's surroundings, and they may please or displease, evoke positive or negative feelings, encourage it to move towards or to move away from the object.

The arguments for a neurobiologically based view of rationality relates to an embrace of such ideas about the development of cognition from an evolutionary perspective, and, in the post-Freudian world, knowledge of the largely unconscious nature of our thinking, and how the unconscious adapts metaphor. The construction of meaning involves a metaphorical transformation of data. This requires, among other things, the development of an increased memory capacity. Initially this would have been very time-limited and situation-specific, but at some point, the efficiency of the episodic and working memory systems must have led to the rise of an early kind of individuality.

Nietzsche's insights into some of the fundamentals of Darwinian ideas have gained acknowledgement from some eminent scholars, especially when it comes to cultural evolution. One of his important insights was that it cannot be inferred that the cause or origin of something will reflect its eventual utility.[8] Nietzsche's concept of the way drives (*Triebe*) influence behaviour reminds us that 'to live you need energy',[9] and Nietzsche was, at least at one point in the later development of his theories, seeking after Dionysus without metaphysics.[10] Nietzsche put the physiological into art, of which drive and

Rausch were key ingredients.[11] In later writings he suggested that *Rausch* even affects the power of vision such that the whole affective system is excited and enhanced. These ideas, according to the philosopher Julian Young, embrace both the Apollonian and Dionysian of the earlier *Birth of Tragedy*.[12] Silk and Stern refer to these urges as proto-creative processes which are, therefore, physiological and psychological: 'On the psychological level, the Dionysiac and the Apolline are creative human impulses, *Triebe* under which are subsumed modes of perceiving, experiencing, expressing and responding to reality.'[13]

We do not obtain knowledge of the environment passively, but by action and motor involvement with our world. There is a powerful urge towards expression, to make an impression on the environment, to attain an end—a momentum without which there could not have been the 'motion' of the emotion, that prepared state, ready to move, to create, and to mate.

THE ORIGINS OF ART

In order to catch a glimpse of the origins of art, we must step back to a time well before the artefacts which have survived, which our ancestors have left behind, and observe the complex patterns that many animals make to enrich their environment, whether for sexual or other purposes. Often quoted are the beautiful bowers which are a part of the courtship ritual of the bowerbirds. These are constructed by males from many naturally found colourful objects, and are inspected by the females, who visit several before choosing her mate by his bower. The duets of the gibbons are another: 'when the female sings, the male starts to sing. During each build-up phase of the female song, the male stops singing. After the climax of the female song, he resumes his song by uttering a response. This response is characterised by its structural complexity, loudness and associated locomotory displays... [the] duets are antiphonal.'[14]

In his book *The Biology of Art*, Desmond Morris discusses the exhibition of chimpanzee paintings which was held in a London art gallery, giving many examples of monkey and ape art from various investigators. Two chimpanzees produced over 600 paintings. The pictures had certain features in common, such as a tendency to fill in a blank page and not to go outside it, to balance an offset figure, or to mark a central figure, which he suggested portrayed a 'hidden potential for painting and drawing'.[15]

The early ancestral artefacts we have, such as stone hand axes dating from about 1.4 million years ago, hint at a cognitive shift in pattern representation. Although earlier stone hand axes are known to be over two million years old, those of the Acheulean era have distinctive oval and pear shapes and a degree of symmetry, sometimes incorporating fossils; some of these are too large or too small to be useful.[16] Typically these are associated with *Homo erectus*, and suggest early artistic sensitivity. Boyd defines art as cognitive play with patterns, Langer as human feelings expressed through perceptible patterns.[17] The potential for this is clearly evident in animal species distant from us in the evolutionary scale, but could be seen emerging at least a million years ago in the hominid line.

Empathy, Theory of Mind, feeling the emotions of others all signal the evolution of *Homo sympathicus*. Through the actions of mirror neurons, humans have the capacity to experience, albeit at second hand, the same 'feelings' as others have, the same knowledge of suffering, a process which Lakoff and Johnson refer to as empathic projection.[18] The importance of gesture and motion to relay emotion, is underacknowledged, but important in the art of Tragic representation. The French philosopher Denis Diderot (1713–84) told of the night he went to the theatre to study gesture, and deliberately sat in the third-class boxes. When the curtain rose he placed his fingers in his ears to obliterate the sound of the actors' voices. He noted how people around him were astonished when 'they saw me shed tears in the pathetic parts, and that with my ears continually stopped'.[19] 'To begin to act as if one had actually entered into another body, another character. This stands at the beginning of development of drama.'[20] Tragedy as an art form surely emerged from reflections of the crystals of tragedy, at a time of the dawn of self-consciousness, the awareness of the I in 'I am'. The Apollonian was always in danger of force changing form, of dissolution and death. Tragedy may be about the death of Dionysus, but tragedy is about the death of Apollo, the individual crafted consciousness.

BACK TO TRAGEDY

This book is about Tragedy, tragedy and emotional crying, a uniquely human response. As we have seen, laughing and crying have intimate social functions, and are crucial in infant–maternal and, later, developing peer-group social communication. There are some distinct differences, however. Crying aloud is present at birth, but crying with tears is seen at around four months, as is laughter. Crying occurs from the self alone, while laughter nearly always happens in interaction with others, especially between mothers and infants early on. Tears communicate intense emotional states quite directly and explicitly, and lead to a response in those witnessing the crying and, in a return response to that, a change in the behaviour and disposition of the person crying.

The studies of adults reviewed in Chapter 2 have shown that the immediate physiological responses to emotional crying are those of activation, and not those reflecting decreased arousal or tension. If anything, the physiology therefore points to an excited affect even though many self-reports suggest that people feel better after crying—although it has been noted that the neurophysiological studies of the after-effects are very limited. The feelings which are retained after the act are significant, like the pleasures of the lingering flavours of an interesting food on the palate.

The self-reports of mood change after crying suggest that to some extent they are dependent on how the episode of crying turned out and who else was present. The psychologist Stanley Schachter conducted experiments which analysed subjects' emotional responses while observing different social situations, to injections of adrenaline or placebo. The physiological arousal to the adrenaline was non-specific, while the cognitive tag given by the social cues enhanced the experiences as specific. Schachter concluded that emotions are very context-dependent, and went on to propose a two-factor theory of emotion which emphasized the physiological as well as the cognitive attributes of the experienced emotion.[21] Human tears require both, but not necessarily in equal proportions.

EYEING

At some time in primate evolution, the importance of the eyes in communication magnified hugely. In most non-human primates, to stare into the eyes of a conspecific imparts an aggressive or a dominance signal, and is best avoided by many members of the troop. Chimpanzee infants and mothers rarely gaze into each other's eyes. Gaze-following, that is, the ability to look in the direction of another's gaze, has been shown in some monkeys and apes, in whom it is more likely to follow head-turning. A human baby's eyes can focus at a close distance shortly after birth, and eye contact between the neonate and mother occurs during breast-feeding. The human infant responds directly to another's eye gaze from around three months.[22] In monkeys and apes the sclera, that is, the protective outer coat of the eye, is pigmented. Human eyes, in contrast, are surrounded by white sclera ('the white of the eyes'), and the colour of the iris is profiled in a soft surround: the pupil expands and contracts to reveal our emotional state.[23] Eye-to-eye contact is discussed by writers and poets of all generations, especially the sorrow and compassion expressed by the eyes, which are often referred to as an entrance to the soul. Ben Jonson's (1572–1637) lines 'Drink to me only with thine eyes / And I will pledge with mine' is half of the story; the other half is from Proust: 'I looked at her, at first with the sort of gaze that is not merely the messenger of the eyes, but a window at which all the senses lean out, anxious and petrified, a gaze that would like to touch the body it is looking at, capture it, take it away and the soul along with it.'[24]

Although reading empathetic signals from eye contact is an essential aspect of human communication, the pupil of the eye, with its surrounding playful iris, may be even more important. In a series of studies, the neuropsychiatrist Hugo Critchley and his colleagues have examined the empathic and cerebral responses to alteration of pupil size. In volunteers, they had the subjects view pupils of differing size, while undergoing fMRI. The subject's pupil size was also monitored. Using the paradigm of viewing happy and sad faces, they reported that looking at small pupil size made sadder faces seem even sadder, but pupil size did not seem to alter the interpretation of neutral, happy, or angry expressions. Also, when the volunteers looked at sad faces their own pupil size mirrored the pupil size observed. This effect was linked with ƒMRI activations of the amygdala and some brainstem nuclei which regulate pupillary size. The brain areas activated when subjects looked at the smaller pupils combined with sad faces included, in addition to the left amygdala, the left insula, the right anterior cingulate area, and some other areas of the temporal and medial frontal cortices, all of which are implicated in the processing of emotional stimuli. These authors conclude that observing pupil size is an important part of emotional social interaction, and it seems closely related to feelings of sadness. Recognizing that those who score highly in emotional empathy are also more sensitive than those with lower scores to detecting subliminal emotional facial expressions, Critchley and his colleagues also measured their subjects' empathy and found that emotional sensitivity to pupil size while viewing sad faces was a good predictor of empathy scores. Thus, pupillary signals are continuously monitored during social interactions, and convey emotional information. Critchley and his colleagues speculate that 'the communication of sadness intensity through pupil size may have evolved to act over short distances, in contrast to emotions such as anger, which may lead to the "withdrawal" of the observing party'. If so, the findings must be relevant to emotional bonding between lovers, as well as between mother and child.[25]

The human infant indulges in prolonged face-to-face interactions with its mother, as it begins to respond to and, as Merleau-Ponty put it, to live in the facial expressions of the other.[26] The valence of the eye for emotional communication is enhanced, as is the value of the emotional significance of crying. The tears first flood the surface of the eye, blurring its appearance, capturing for the looker the first seconds of the emotional turmoil and potential torrent which is about to begin, around the edges of the eyes, down the cheeks and side of the nose, to flow and fall even as far as the floor. Emotional crying is a human communication revealing suffering and pleading for nurture and help, and the benefits for the growing infant are clear. The infant eye and cry brings the mother close, with her caresses. The patterns of such early infant behaviour endure, but in adulthood become more controlled, although they can be released by appropriate stimuli at any time in life.

The rewarding effects evoked by tears reinforce the behaviour. Experiences that acquire emotional connotations during memorization will, with re-membering on recall, be impregnated with that emotion, and the metaphorical identifications of the memory. The latter include, after a certain age, the identification not only of bonding, but also of loss, separation, and the fear of annihilation.

In Wagner's opera *Tristan and Isolde*, Isolde recounts to her maid Brangäne how at an earlier time she had healed a sick man called Tantris, who had travelled to Ireland to seek her healing powers. After treating him, she realized that he was none other than Tristan, the knight who had slain her betrothed, Morold. She was moved to kill him, as he was an enemy to both her and Ireland. He looked into her eyes, and instead of striking him the sword fell from her hands, and she was unable to harm him. In the opera, when they have both drunk what they supposed to be a death potion, but which has been exchanged by her maid for a love potion (which even so signals the eventual deaths of the lovers at the end of the opera), they gaze fixedly into each other's eyes, to glorious music, enraptured.[27] As Scruton puts it, 'The text and the music deftly remind us of a singular fact: that we look *at* inanimate objects, and we look *at* human limbs, but we look *into* someone's eyes, and every such look is compromising, fraught with significance, a face-to-face encounter with the other, and therefore a summons to hate or to love.'[28]

It is interesting to speculate further on the importance of human sexual attraction. Thus, at one point in primate evolution, face-to-face encounters became the anatomically and socially preferred way of copulation, as sexual dimorphism in body size decreased and pair bonding increased. The longer-legged, pink-lipped bonobo has been observed to engage in face-to-face sexual intercourse and tongue kissing.[29] With its notable facial expression, it is thought to have separated evolutionarily from the common chimpanzee some one million years ago. It walks upright at least some of the time, and is thought to mate throughout the year, its sexual activity being less bound up with the female menstrual cycle. The power of the orgasm, the facial expressions, and the closeness of the eyes between the couple must have further enhanced bonding and the importance of the eyes as windows of emotion-perhaps even the beginnings of the emotion of love.

THE NURTURE OF NATURE

It is unclear when emotional crying emerged on the evolutionary stage, but the close biological interactions between the basic, mainly pre-set neurological underpinnings of the emotions and cultural influences are now much clearer. As Damasio suggests, the cultural modifications 'first shape what constitutes an adequate inducer of a given emotion; second, they shape some aspects of the expression of the emotion; and third, they shape the cognition and behaviour which follows the deployment of an emotion'. The plasticity of the human nervous system has allowed all objects to be endowed with emotional value for us, some more than others. For circumstances to trigger feelings, the signals given by the inducer of the emotion activate the basic neural circuits already discussed (amygdala, prefrontal cortex, brainstem autonomic nuclei), which widely distribute activity to other brain areas (cortical and subcortical), and the body (the viscera especially). This causes what Damasio refers to as the whole commotion, the emotion as a neural 'object'. This has both somatic and cognitive components, relating to the speed and focus of sensory processing and onset of specific behaviours, including those aimed at bonding and nurturing. The evolutionary significance of the emotions is profound: they 'are not a dispensable luxury...they are part and parcel of the machinery with which organisms regulate survival'.[30]

Nietzsche referred to the *pudenda origo*, the shameful origins of our cultural beliefs and practices; the evolutionary backdrop to our emotions and behaviours cannot be ignored. Protected by our cosy 21st-century arrogance towards all that is bygone, we have swerved off at a tangent from wanting to acknowledge or accept our biological heritage, and lost contact, emotionally and intellectually, not only with the myths of our forefathers, but with our attachment to the natural and the animal world from which we have evolved. We have been unhinged from our identification with 'the primordial being' itself, to use Nietzsche's own expression (*das urwesen Selbst*). As the last ice age began to release its hold on the world's continents some 22,000 years ago, the tundra in Europe receded, global warming ensued, and our ancestors and their arts flourished. In response to local climate fluctuations and seasonal changes, food was stored but needed to be protected from others; rights and rituals arose; and what is referred to as civilization developed. The survival of communities, which now depended on internal cooperation, must have increased the need for more precise means of communication. This led to the development of early efficient spoken languages, which advantaged some groups above others. Bonding between individuals was an essential aspect of this.

It should be recalled that the title of Darwin's book on natural selection was *On the Origin of Species by Means of Natural Selection; or The Preservation of Favoured Races in the Struggle for Life.*[31] Existence involves individuals struggling with their environment and with each other, the struggle being most severe between individuals of the same species in a world where Malthusian principles apply—that is, where more individuals are born than can survive on the limited food sup-ply.[32] In his later book *The Descent of Man*, Darwin, however, highlights social instincts and cooperation, and states: 'the more important elements for us are love, and the distinct emotion of sympathy'.[33]

Crying with tears is an expression of a non-propositional category, which emerges in early infancy, and is closely inter-woven with infant–maternal bonding. As an individual matures, crying tears remains and persists as one of the most powerful of social signals. Since the origin of some body part or of a behaviour and its ultimate utility are not closely bound in evolutionary time, the origin of tears, their link with emotions, and their value in human communication need not coincide. Tears have value way down the evolution-ary scale, associated with protection of the eyes from drying, and/or with antiseptic properties from infection, but at some point over the millennia, they acquired some new proper-ties, and different meanings and values became attached to them.[34]

As Darwin stated: 'We may infer as highly probable that we ourselves have acquired the habit of contracting the muscles round the eyes, whilst crying gently, that is without the utter-ance of any loud sound, from our progenitors, especially dur-ing infancy, having experienced, during the act of screaming, an uncomfortable sensation in their eyeballs.'[35] For Darwin, actions which were at first voluntary became habitual and then hereditary, tears being a reflex action from spasmodic contraction of the eyelids and surrounding muscles, and when the eyeballs become gorged (with blood). In his view such tearing could not have appeared in evolution until the muscles around the eyes had acquired their present structure—with the development, therefore, of our complex social engagement system: the expression of grief with tears is evidently a late hominid attribution. Suffering readily causes the secretion of tears, and the secretion of tears serves as a relief to suffering, as crying is abated by amelioration of the state that led to the crying.[36]

It will be recalled from Chapter 3 that afferent input from the facial muscles and skin directly influences neuronal activity in brain circuits that are linked with emotion. It has also been shown that viewing an emotion can induce subtle changes in the muscular activity of the face of the viewer, even if the stimulus is not consciously registered, the activity being detected only by measuring the muscle activity with electromyography.[37] This can occur after viewing video clips of synthesized emotions, and the responses do not habituate with repetitive presentations.[38]

Thus from an anatomical perspective, observing an emotion in another leads to changes in the muscle activity around the eyes, which can stimulate a reflex through the cranial nerves to the lachrymal gland provoking tears. The response is, however, influenced by other changes induced in central cerebral structures, which include those of the mirror neuron system, and their association with the circuits that relate to empathy and 'feeling'. The latter, in terms of Damasio's markers, will magnify with the involvement of peripheral visceral influences, which, via the vagus inputs to the autonomic system, are part of the cascade of emotion, which, in the appropriate setting, and permeated by personal memory, leads to emotional crying, tearing, and the associated gut feelings.

With maturation, top-down influences from the higher cortical structures come to predominate, conditioned by individual social circumstances. If it is the case, as suggested by the research discussed earlier, that people with a greater tendency to empathy are the more inclined to imitate emotion, then they may be the more inclined to extra-conscious mimicry and to possess a greater facility to cry in response to what they see in the face and eyes of another. Seeing sadness or witnessing crying provokes the initial stages of facial muscle contraction, leading to the chain of events just described. The main cortical components involved in these events are the insula, the cingulate gyrus, the medial and orbital frontal cortices, the dorsal frontal cortex, and the amygdala.

7

WHY DO WE GET PLEASURE FROM CRYING AT THE THEATRE?

In ancient times a story could only end in two ways: having passed all the tests, the hero and the heroine married, or else they died. The ultimate meaning to which all stories refer has two faces: the continuity of life, the inevitability of death.

Italo Calvino, *If on a Winter's Night a Traveller*[1]

Whether or not there is a human experience that may be specifically referred to as aesthetic is a philosophical quandary. As we saw in Chapter 5, several scientific researchers have attempted classifications of the emotions, but others, mainly non-scientists such as Langer, have pointed out that our vocabulary is simply not adequate to parcel out the fine spectrum of our feelings into discrete categories. What we seek to express is not expressible in ordinary discourse, which, as Langer remarked, is 'peculiarly unable to articulate the character of our so-called "inner life"'.[2] Apollo cannot deconstruct Dionysus.

I have put forward the view that there is a special feeling associated with Tragedy, which has been embraced under several names, but which in Chapter 5 I referred to as tragic joy. The emotions evoked by the art of Tragedy, crystallized in drama, are viewed in this book as emerging from the cognitions of our ancestors, at the dawning of self-conscious individuation. Among theoreticians, the potential to stimulate the aesthetic emotion is often argued to be a property of the object that excites the emotion, with form dominating over content. However, this suspends the emotion itself in limbo, shifting the focus away from the aroused feelings and the close bond between feelings and meanings. The construction of meaning is not simple information processing: metaphorical transformation of data is the important cognitive step. The aroused aesthetic state is an active one physiologically, not a catharsis, and not, as was argued by some philosophers circling around the ideas of Kant, one of passive disinterested contemplation.[3] The viewer is actively involved.

The neuroanatomical sections of this book have described the ascending and descending streams of neural activity that occur with the flow of emotional tears, and the events that are most likely to provoke them have been discussed in Chapter 2. Many people 'know' the feeling of tragic joy, and have experienced this in the aftermath of crying. Recently, further neuroanatomical indicators as to the nature of this special feeling have been revealed.

Recall Aristotle's prime emotions stirred by Tragedy, namely pity and fear, and the role of the amygdala in the generation of emotional responses. To repeat, the amygdala is a central anatomical component in the circuitry of the emotions, with extensive cortical and subcortical connections, activated by a range of emotional stimuli, and linked to their interpretation. Through association learning it provides positive or negative valuation to stimuli. While much of the experimental work has involved the negative emotions, especially fear—using the fearful faces paradigm, in which responses to the presentation of fearful and neutral faces are compared using fMRI—it is also linked with positive emotions.

The amygdala responds to a spectrum of emotional faces, including those depicting sadness.[4] Neuroimaging studies of normal volunteers who have been asked to simulate emotional states have compared those induced by happiness with those of sadness. With regards to sadness, although there is not complete uniformity between the studies, cortical activations have been seen in several structures, already discussed in relation to emotionality (the insula, anterior cingulate, medial and orbito-frontal cortex), whereas in the state of happiness reductions in cortical activity have been reported. Mark George and his colleagues point out that these differences are reflected in other studies in which the euphoria induced by taking morphine or cocaine is also associated with decreases in regional brain activity in frontal and temporal areas. Other studies, in which subjects have reappraised a fear stimulus, reinterpreting its emotional significance with a reduction of the negative affect, have also shown a *decrease* in activity of the orbito-frontal cortex and amygdala, suggesting a top-down modulation of the meaning of the stimulus, probably from areas of frontal cortex.[5]

A key anatomical distinction of the tragic emotion (tragic joy), which distinguishes it from fear, I suggest, is *decreased* activity of the amygdala. There is evidence that supports this possibility. The neurobiologist Semir-Zeki, in his analysis of the *Splendors and Miseries of the Brain*, looked at the neural correlates of romantic love. Noting that such emotions are often triggered by visual cues, he compared the brain activity of volunteers when looking at images of the one they love as opposed to friends. Areas of increased activity were shown in the insula, anterior cingulate and hippocampus, and components of the reward system—the subcortical ventral striatum. But there was *decreased* activity in some cortical areas, including the lateral prefrontal cortex, and in the amygdala. Similar findings in the amygdala applied when maternal love was examined, mothers being shown images of their own as opposed to other children.[6]

In a more recent paper entitled 'Toward a Brain-Based Theory of Beauty', Zeki looked at the ƒMRI responses of volunteers to the experience of beauty, when looking at paintings or listening to musical extracts which had been graded by the subjects into categories of 'beauty', 'indifferent', or 'ugly'. The experience of 'beauty' for both art forms activated the orbito-frontal cortex, while 'ugliness' activated the amygdala, this structure not being perturbed by contemplating 'beauty'. While these studies have been concerned with emotions linked with love, maternal care, and beauty, they support a view that amygdala activation, as seen in relation to fear, is not the neurophysiological basis for these tender emotional states.[7]

A further link between the activity of the amygdala and a calming effect on emotion comes from a treatment used in epilepsy and also chronic depression, namely vagus nerve stimulation. The importance of the vagus nerve and its nuclei for modulating emotional activity was discussed in Chapter 3, where it was noted how there are direct inputs from the solitary nucleus to the amygdala. Studies with brain imaging have been carried out examining the effect of vagus stimulation, albeit in people with severe neurological or psychiatric illnesses. The data are difficult to interpret because of the different conditions of the patients treated, the different times of investigation, and the different methods. However, given these caveats, one effect of stimulation of the vagus nerve input is to decrease activity in some structures, including the cingulate gyrus and the amygdala.[8] Although it is by no means certain, one interpretation of the findings is that reduced amygdala responsiveness is associated with the antidepressant effect of vagus nerve stimulation.

The implication of these studies is that the feelings aroused by Tragedy, which I suggest are different from those that are usually listed as typical human emotions, are different because of their neuroanatomical and evolutionary bases. The underlying neurology is not that of evoked fear, but is akin to feelings associated with social bonding and love, and similar to those linked to our aesthetic appreciation of the beautiful. The neurological components of this feeling hinge on the amygdala, in its interplay with other brain and bodily structures, that resonate with our emotional experiences. These do not, from a physiological perspective, lead to a catharsis, but to a sensation of intimacy which is a special combination of arousal and calm.

DANCING IN TIME WITH THE MUSIC

Nietzsche considered Tragic pathos to be closely related to feelings aroused by music.[9] For him, the birth of Tragedy emerged from the spirit of music. The responses of limbic structures to music support this association. The word 'melody' (Greek, *melōidia*) derives from *melos* (tune) and *ote*—words that refer to our ability to hear changes in pitch and to sing to them, such singing synchronized to rhythm, the ability to entrain to rhythm being another unique human attribute. The first musical instrument was the voice, and the history of modern Western music began with an emphasis on the word, as sung in medieval plainchant. With the rise of opera from about 1600, initially with the idea of recreating the essence of Greek theatre, music began more and more to emphasize the emotional intensity of the words, and hence the drama, which gave rise to a need for new musical instruments to expand the expressive meaning of the words. With the development of instrumentation, instrumental music without words became fashionable, as it was readily accessible to all, heard in concert houses, or as chamber music in intimate settings such as the home. The major triad was the foundation stone of the Western harmonic system, triadic harmony having dominated our music for over 500 years.[10] However, according to Philip Ball, the scales of ancient Greek music were composed of sounds similar to those in use today which are referred to as the diatonic scale, that is, proceeding by whole tones. If this is so, the music we listen to is based on very old tonal appreciation—that is, music with a tune, with a tonal system anchored to tonal centres possessing tonic gravity.

The tonic pitch on which harmonies are built (say the note C), offers chords which, by means of progression from chord to chord, using such musical techniques of composition as repetition, modulation, and transformation, move away from these centres only to return with harmonic resolution. Scruton uses the term 'acousamatic' to define what we hear in music: not just a succession of sounds, but 'a movement between tones, governed by a virtual causality that resides in the musical line'.[11] Within the music phrases, calm and tension are developed, with dissonant chords urging a return to the tonic. The discord requires the concord, since it provokes restlessness, suspensions, and anticipations all requiring resolution—in such music there is not only melodic, but also harmonic and rhythmic closure. We are home, as Anthony Storr put it, but not until our expectations have been confounded on a journey of contrast, conflict and delayed resolution.[12] To quote Scruton again: 'We cannot hear musical movement without seeking for points of stability and closure—points towards which the movement is tending or from which it is diverging, and to which it might at some point "come home"…there are a priori constraints on musical syntax.'[13]

The neuroanatomical responses on listening to music support the importance of the amygdala in the experienced affect. The studies have yet to reveal any cultural differences. They consistently report increased activity in response to emotionally pleasing music in the ventral striatum—the basal ganglia structure so involved with motivation, reward, and pleasure. With regards to the amygdala, Robert Zatorre and colleagues used PET imaging to compare the response at moments of 'shivers down the spine' or 'chills', when subjects were asked to listen to pleasant as opposed to unpleasant music.[14] Increasing intensity of chills led to increased activity in several areas including the insula, ventral striatum, and anterior cingulate cortex, but *decreases* in the right and left amygdala, ventromedial prefrontal cortex, and precuneus. Chills were associated with physiological responses of increased heart rate, muscle activity, and respiration rate.

Using an *f*MRI paradigm, Stephan Koelsch and his colleagues have shown a similar effect on the amygdala, with *deactivations* in response to pleasant musical stimuli.[15] Zatorre states: 'Thus, activation of the reward system by music may maximize pleasure, not only by activating the reward system but also by simultaneously decreasing activity in brain structures associated with negative emotions.'[16]

Further, the frontal cortex responds to music. Petr Janata and his colleagues have reported that the medial prefrontal cortex has been shown to track the movement of a melody through tonal space created by the system of major and minor keys underlying Western tonal music. Further, they elicited autobiographical memories with excerpts of popular music dating from an individual's past, and showed that the dorsal medial prefrontal cortex was activated when experiencing emotionally salient episodic memories triggered by familiar songs. Emotional responses to music and the perceived pleas-antness of music have also been shown to modulate activity in the ventral medial prefrontal cortex when short consonant musical passages were compared with the same passages rendered increasingly more dissonant.[17]

Music moves in time. Music has nothing to do with rational agency, but it re-sounds through the nervous system in ways which we can now discern anatomically. The responses are non-cognitive, non-rational, and inescapable. As John Ham-ilton has remarked: 'music tracks down everyone within its range, rudely and without regard. Unlike the eyes, the ears have no lids.'[18] Langer suggests that the emotions felt in response to music are different from ones we normally know, and the musicologist Deryck Cooke referred to musical emo-tions being 'a different kind of feeling...the same but also something other'. Both commentators thus echoed the above suggestions with regard to emotions evoked by Tragedy being special, and re-emphasize the close link between music and Tragedy.[19] The manipulation of expectation in music has been commented on already—how the listener is variously led towards and away from the final tonic resolution, and how the amygdala and its surrounding structures respond with these yearnings, earning for the listener the pleasure of reward and composure.

There is in Nietzsche's philosophy a connection between art and suffering. The tragic hero, human and ephemeral, is destroyed by the greater forces that encase his life; the Dionysian always dominates, Dionysus' death being a metaphysical metaphor of becoming. Discussions of the meanings behind Tragedy circle around life, death, and love. Tragedy as art form hijacks and crystallizes primordial feelings and emotions, which affect and unite us most deeply as individuals embedded in the tragedies of our everyday lives. The words and music of Attic tragedy live on, from the chorus to contemporary cinema.

While events surrounding birth and death shadow so much of our personal lives and our dramas, in much Tragedy and tragedy it is love which binds together the plots of our lives and plays.

AND WHY DO ONLY HUMANS CRY?

And now abideth faith, hope, and love, even these three: but the chiefest of these is love.

(Geneva Bible, 1560)

It is my belief that two facets of the human mind that are universal, and have been so important in driving the development of human cultures, are hope and compassion.[20] The evolution of empathy, arising from a Theory of Mind, mirror neurons, memory, our ability to foresee a future, and the neuroanatomical accompaniments to these gave us a brain that responds with compassion; we feel suffering.These have promoted behaviours of profound significance for the individual and driven the development of culture. Although compassion is not only a human attribute, its hypertrophy in humans has given us the potential for social interactions of the most sensitive and profound nature. The human capacity for empathy, like other behavioural traits, is not present to the same degree in everyone, some people apparently, as Simon Baron-Cohen puts it, possessing zero degrees. He argues that empathy is universal in human cultures, is measurable, and is distributed in a bell-shaped curve, such that some people have no empathy at all. His models have been people with autism and autistic-spectrum disorders, and the use of Theory of Mind tests in his studies. For the underlying neuroanatomy, he relies on the same brain circuits as described in this book. He even argues for empathy genes, which have been investigated by him and his colleagues.[21] But feelings of empathy engendered by seeing the faces of suffering, have been enhanced at some time in hominid development by the signal of tears, providing a boost to empathic identification and all that went with it, including our emotion of love.

Tears are an accompaniment to Tragedy as art form, and they reflect the tears of everyday human tragedy, which is linked to loss and mourning. These feelings have arisen in the course of our long evolutionary history, notably with the rise of self-consciousness, the development of small communities, and the growing potential for love and hence an even greater sense of loss. Binding these together is music, the traditional art form that most moves us to tears, and without which many social occasions would be bereft of their meaning and pleasure. Since our response to crying in many settings turns out to be rewarding, which is again linked to the underlying neurobiology of the emotions, the arts that evoke such emotions have flourished, an ancient echo from our ancestral past.

Neuroanatomy

Neuroanatomy is complex. This is because, to understand it, the brain has to be viewed in three dimensions, an unusual language has to be learned, and the complexity of different connections and circuits renders appreciation difficult. This is very daunting for the initiated and the uninitiated. There is a tendency after a while to think that everything in the brain is connected to everything else, and that any hypothesis can be verified by simply implying that two structures must connect. The brain simply does not work like that, and nuclear clusters (nuclei) and their connecting fibres (tracts) are beautifully arranged, and have been patterned with precision over aeons.

This summary of the main neuroanatomy described in Chapters 3 and 4 is intended to help the reader, whatever their pre-existing neuroanatomical knowledge, through the essential details in the kindest way possible. The reader is referred to the relevant illustrations in Chapter 3 to help with this text. It begins at the top of the brain, and works down.

When viewed from above, the **cerebral cortex**[1] of the brain is seen. It is composed of **neurons** or nerve cells and their fibre extensions, the **dendrites** and **axons**. It is by the axons that neurones interconnect, and it is at the dendrites where one neurone will, by the interplay of **neurotransmitters**, trade information with others. The cortex is referred to as neocortex, as it is quite new in evolutionary terms, and on microscopic inspection it is observed to be laminated and six clearly visible layers are seen.

The cortex is divided into lobes: the frontal, parietal, occipital, and temporal lobes. For the purposes of this book, the most important are the frontal and temporal lobes. The most forward part of the frontal lobe is the prefrontal cortex, which has been divided into several subregions: lateral, medial, and orbital (situated just above the eye). The medial prefrontal cortex plays an important role in this book, being related to several aspects of the **social brain**. The medial frontal cortex is continuous with the **cingulate gyrus**, an elongated structure extending backwards from the frontal regions of the brain, and then looping down posteriorly to link with structures in the temporal lobes.

Underneath the cortex lie many nuclei, some very small, others forming large clusters readily visible to the eye. The **basal ganglia** are a group that receives information from the cortex, and form part of a system for regulating motion and motivation. The most important structure relating to emotion is the **striatum**, a part of the basal ganglia, especially that part under the head of the striatum called the **ventral** striatum. The **thalamus** is a collection of nuclei that interweave their activity with the basal ganglia, ensuring coordination of motor activity, but the thalamus is also a major byway station of sensory information travelling from sensory receptors at the surface of the body (eye, skin, etc.) and the interior, such as from the gut, heart, and lungs to the cortex.

The **limbic system** is the evolutionary older cortex. It is laminated, but not in six layers, much of it being two or three layers. It has been linked to emotion, and has several important components. Most relevant are the **amygdala** and the **hippocampus**. These lie close to each other in the temporal lobes, and influence each other, but have extensive important outputs to the cerebral cortex and the basal ganglia. The cortical outflow alters the tone of sensory information coming into the cortex from sensory receptors and the thalamus (giving information about the environment). The outputs to the basal ganglia provide information about the emotional state and highlight the importance of the immediate sensory perceptions. It is the outputs from these medial temporal structures to the ventral striatum that are most significant, the activity of the ventral striatum being associated with motivation—a limbic–motor interface. The information flow around the basal ganglia, thalamus, and cerebral cortex is thus in part coordinated and contextualized by the limbic, especially amygdala, input. Lying over the hippocampus is the **parahippocampal gyrus**, which connects with the tail end of the cingulate gyrus, and provides information to the hippocampus from the cerebral cortex.

Another important limbic structure is the **insula**. Unlike most of the limbic structures which are best visualized from the medial surface of the divided brain, the insula is buried beneath the cerebral cortex, and is seen, if the cortex is removed, from the lateral side. It is the part of the cortex which receives much information from the viscera (guts).

The amygdala is connected to many other areas, and in terms of the neuroanatomy of crying, its inputs to the orbito-frontal cortex, **hypothalamus,** and **brainstem** structures that regulate **autonomic activity** are important. The hypothalamus is a nuclear cluster situated close to the centre of the brain, adjacent to the **ventricular** cavities which contain the **cerebrospinal fluid**. It regulates the output of hormones, via the pituitary gland situated beneath it, but its outflow down to the brainstem carries information from the orbito-frontal cortex, amygdala, and hippocampus.

The brainstem contains many nuclei, such as those that regulate the **cranial nerves**, which exit from the skull and control the activity of, for example, the facial muscles and eye movements. It contains fibres of passage travelling from the cerebral cortex to the spinal cord which influence the final output of the motor nerves controlling muscle activity, and those tracts passing up from peripheral receptors to the thalamus.

Cranial nerve X is called the vagus nerve. It is regulated by three intertwined brainstem nuclei called the dorsal motor nucleus of the vagus, the nucleus ambiguus, and the nucleus of the solitary tract (solitary nucleus). Information from much of the interior of the body, such as from the gastrointestinal tract, enters the brainstem to terminate in the nucleus of the solitary tract, and from there is linked by direct pathways with the amygdala. There is reciprocal influence, since the amygdala plays back to the vagus nerve complex, which also receives direct downward connections from the insula, hypothalamus, and the prefrontal, especially orbito-frontal, cortex.

Other important links are those from the solitary nucleus to the insula, and from the insula to the amygdala and the prefrontal cortex.

The act of tearing requires output from the lachrymal nucleus situated just above the nucleus ambiguus. This is achieved through cranial nerves which supply the lachrymal glands around the eye. The lachrymal nucleus is influenced by the downward connections as discussed above.

The Human Brain

This book is concerned with emotional crying and why this is a unique human attribute. One way to answer this question is to seek differences between the human brain and that of other primates. Five principles are relied on.

1. The limbic system is closely integrated with the basal ganglia structures that refine and regulate our motion and emotion.

2. The limbic system is closely integrated with the cerebral cortex, and influences higher cognitive activities, including thinking and decision making.

3. The primate brain, which evolved into the human brain has increased in size greatly over a few million years, especially those areas relating to memory, language and the social brain.

4. In the human brain, output from the cerebral cortex to the brainstem structures regulating the autonomic nervous system is direct, whereas in other primates it is largely indirect.

5. The complex social life of humans has underlying drives, understandable with evolutionary and neuroanatomical principles, which explain the origin of many cultural activities, especially creativity and our appreciation of art.

GLOSSARY OF TERMS

Amygdala a nuclear cluster in the medial temporal lobe, lying next to the *hippocampus*, associated with emotional memory, and *valence*.

Autobiographical memory memory that is related to personal life events.

Autonomic nervous system the brain system that regulates autonomic functions. These relate to many body activities, such as breathing, heart rate, and the activity of the gastrointestinal tract, and are largely outside conscious control.

Axon the long tube of the neuron that delivers neurochemicals from the nucleus to the *synapse*.

Basal ganglia a collection of nuclei situated under the cortex, and close to the *thalamus* which link with the cortex and the thalamus to regulate motor activity. The main nuclei are the caudate, putamen, and globus pallidus. The head of the caudate contains the *ventral striatum*, which is closely linked to the amygdala.

Brainstem the part of the brain situated between the mid-brain and the spinal cord. Two sections are referred to as the pons and, beneath it, the medulla.

Broca's area in conventional neuroanatomy this is the area of cortex situated in the left lateral inferior frontal region (Brodmann areas 44 and 45). Damage here leads to aphasia, the loss of propositional speech. The homologue in the right hemisphere is also important in an understanding of the social brain, but is not usually named after Broca.

Cerebrospinal fluid the fluid which surrounds the brain and spinal cord.

Cingulate gyrus a long, curved C-shaped gyrus which is part of the limbic system.

Cortex the surface of the brain, some of which can be seen with the naked eye, but much of which is hidden in the *sulci*. It is laminated, and the number of layers varies from two to six.

Cranial nerves these are numbered 1–12, and leave the brain via the skull. Of most importance to the text is cranial nerve X, the vagus nerve.

Dendrite small extensions of the neuron which make contact with *axons*, and where many of the brain's synapses occur.

Dorsal on the back of, above.

Electroencephalography measurement of the brains electrical potentials (brain waves) using scalp electrodes.

Electromyography measurement of muscle activity.

Genotype the genetic make-up of an animal or plant.

Glia part of the structural apparatus of the brain. There are many more glial cells in the brain than there are neurones. They are referred to by names such as astrocytes. Their function is still poorly understood.

Gyrus the structures that give the surface of the brain its smooth, undulant, sculpted shape. They are composed of neurons, glial cells, axons, and blood vessels.

Hamartia a flaw or an error, usually applied to a person or character. In the context of Tragedy it referred to a weakness or failing in the tragic hero, which brought about their downfall.

Hippocampus a part of the limbic system, at the front of the medial temporal lobe, lying adjacent to the amygdala, and extending backwards to curve round and meet with a part of the tail of the cingulate gyrus. It is covered by the *parahippocampl gyrus*, and is linked to the laying down of day-to-day memories and their recall.

Hominid this usually refers to humans, and their ancestors (who are all extinct!).

Hypothalamus a collection of subcortical nuclei which regulate much *autonomic* and hormonal activity. It is situated medially, adjacent to a *ventricle*.

Insula a part of the limbic system which, unlike the other limbic structures, cannot be seen from the medial side, and is buried under the cerebral cortex from a lateral view. It is the cortex which receives *visceral* information.

Lachrymal pertaining to tears.

Limbic system over time the structures described as limbic have varied considerably. The limbic system is identified with circuits that mediate emotion. One definition is the *amygdala*, *hippocampus*, and their direct connections. The definition used in this book is all *cortex* that is not six-layered. This identifies the limbic system as older cortex, from an evolutionary perspective.

Logos a word with many nuances, but closely linked to ideas of structure, reasoned thought, and rationality. For Aristotle, humans were distinguished from other animals by their possession of logos.

Mental time travel the ability to imagine the future.

Neocortex literally, cortex which is new. Cortical structures can be found in the brains of many animals, not only mammals.

But it is the neocortex which has expanded so magnificently in the human brain to allow for much cultural development including language. It is six-layered.

Neotony the biological and evolutionary process which has led to the prolonged period of dependency of the infant and growing child on a carer or carers before achieving independence.

Neuron the main structural cell of the brain.

Neurotransmitter the chemicals which are released at synapses to influence activity of the post-synaptic *dendrites*. Well-known names include serotonin and dopamine.

Ontology to do with the nature of being and becoming.

Parahippocampal gyrus a strip of limbic cortex overlying the hippocampus. It is a major relay station for passing cortical information to the hippocampus and vice versa.

Phenotype the characteristics of an individual, in part related to the underlying *genotype*, modified by epigenetic and environmental factors.

Precuneus a part of the parietal cortex situated posteriorly and medially. Recent studies suggest a central role for the precuneus in a wide spectrum of highly integrated tasks, including visuospatial imagery, episodic memory retrieval, and self-processing operations, namely first-person perspective taking and an experience of agency.

Principium individuationis a term which refers to the concept of individuation—the boundaries which separate us from the world and from each other.

Prosody the affective aspects of intonation, which gives speech its rhythmic and musical qualities.

Readiness potential an electroencephographically recorded scalp potential seen to arise before a movement is actually made

Social brain the brain structures that seem to relate to higher social activities, such as Theory of Mind and altruism.

Striatum part of the *basal ganglia*.

Sulcus the valleys between the gyrae, where the latter are infolded.

Synapse where neurotransmitters are released from pre-synaptic neurons to influence the post-synaptic neurons.

Thalamus a nuclear cluster situated subcortically, close to the basal ganglia.

Theory of Mind the concept that one person can know that another has a mind like theirs, and that they can influence that mind—and deceive it!

Valence the emotional value associated with a stimulus.

Ventral below.

Ventral striatum the part of the caudate nucleus below the head of the caudate, to be contrasted with the *dorsal* striatum.

Ventricle a fluid-filled space in the brain, surrounded by brain substance. The fluid is *cerebrospinal fluid*.

Visceral relating to the internal organs of the body, especially the guts.

Working memory memory that holds information on line for a brief period of time, such as a telephone number. It is linked to frontal cortical activity.

NOTES

Chapter 1

1. F. Nietzsche, *The Gay Science* (1881), trans. W. Kaufmann, Vintage, New York, 1974, aphorism 342.

2. M. Bekoff, *The Emotional Lives of Animals*, New World Library, Novato, CA, 2007.

3. C. Darwin, *The Expression of the Emotions in Man and Animals* (1872), in *From So Simple a Beginning: The Four Great Books of Charles Darwin*, ed. E. O. Wilson, W. W. Norton, New York, 2006, pp. 1355–8.

4. J. Masson and S. McCarthy, *When Elephants Weep: The Emotional Life of Animals*, Vintage, New York, 1996, p. 133.

5. D. Fossey, *Gorillas in the Mist*, Houghton Mifflin, Boston, 1983, p. 110. An American zoologist who studied gorillas in Rwanda, Fossey was murdered by poachers on 27 December 1985.

6. Jane Goodall studied chimpanzee behaviour for 45 years in the Gombe Stream National Park in Tanzania. The resident primatologist at the Jane Goodall Institute has observed with regard to chimpanzee tears: 'I have only seen them give emotional responses (screaming, whimpering etc.) but no actual tears. I do not believe that chimpanzees have the ability to shed tears—I believe that this is unique to humans amongst the primates.'

7. M. R. Trimble, *The Soul in the Brain: The Cerebral Basis of Language, Art, and Belief*, Johns Hopkins University Press, Baltimore, 2007.

8. M. Scammell, *Koestler: The Indispensable Intellectual*, Faber and Faber, London, 2009, p. 231.

9. Arthur Schopenhauer's use of the term 'Will' is not with the usual meaning of the word. It implies a kind of force, cosmic in dimension and quite impersonal, which has no connection with the mind or consciousness; as Bryan Magee puts it, 'the nearest we can even come to having direct experience of one of its manifestations is our own

acts of will, in which we experience from within the otherwise inexplicable go, drive, force, energy instantiated in physical movements' (*The Story of Philosophy*, Dorling Kindersley, New York, 2001, pp. 140–1).

10. Music for Schopenhauer is the art which does not represent, in contrast to other arts which 'give to airy nothing / A local habitation and a name' (Shakespeare, *A Midsummer Night's Dream*, V. i).

11. A. Schopenhauer, *The World as Will and Idea*, J. M. Dent, London, 1995, pp. 43, 85.

12. F. Nietzsche, *The Birth of Tragedy: Out of the Spirit of Music*, trans. S. Whiteside, Penguin, London, 1993, p. 14.

13. Nietzsche, *Birth of Tragedy*, p. 16. Principium individuationis is a term originally used by Schopenhauer and then by Nietzsche to refer to the concept of individuation. Apollo slayed the snake (python) which guarded the Delphic oracle, and this allowed him to seize the shrine and the oracle for his own worship.

14. The Socratic dialectic was a form of critical oppositional argument, an inductive method to reveal the truth of a proposition, and a way to the logos, a word with many nuances, but akin to logical thought.

15. Nietzsche, *Birth of Tragedy*, p. 66.

16. Nietzsche, *Birth of Tragedy*, p. 89.

17. Nietzsche, *Birth of Tragedy*, pp. 4–5. It seems that when he wrote the preface to the second edition, referred to as 'attempt at self-criticism' (pp. 1–12), he did not have the original text with him (J. Young, *Friedrich Nietzsche: A Philosophical Biography*, Cambridge University Press, Cambridge, 2010, p. 438).

18. Nietzsche, *Birth of Tragedy*, p. 6.

19. Emphasis added.

20. G. W. F. Hegel (1770–1831) was a German philosopher central to German idealism, who developed ideas about the relationship between the universal Spirit and the individual consciousness. His dialectic method, different from that of Socrates, places oppositions against each other (thesis and antithesis), from which synthesis is achieved.

21. The seeds of Nietzsche's synthesis were well sown, but according to Silk and Stern in *Nietzsche on Tragedy*, although the time with Wagner was a stimulus to the development of his ideas, his construct of the relationship between Apollo and Dionysus in the context of Tragedy

was new and very original: 'Nietzsche's Dionysus and Apollo start as historical Greek gods and finish as entities with a supra-historical—or at least "world-historical" character; at a certain point they take on an "artistic" life of their own' (M. Silk and J. P. Stern, *Nietzsche on Tragedy*, Cambridge University Press, Cambridge, 1981, p. 180). Nietzsche sculpted a new theory of art out of these three influences, namely philosophy, music, and the ideas of Wagner. For him, beauty, the high point of Greek art, was insufficient to explain the differences between music and the visual arts (Nietzsche, quoted in G. Liébert, *Nietzsche and Music*, University of Chicago Press, Chicago, 2004, p. 43).

22. *Trieb* can be translated as drive, urge, or instinct in English, and it can be argued that drive and instinct are different phenomena. *Instinkt* (Ger) is another word for instinct.

23. J. Sallis, 'Shining Apollo', in W. Santaniello (ed.), *Nietzsche and the Gods*, SUNY Press, New York, 2001, p. 62.

24. Nietzsche, quoted in G. Moore, *Nietzsche, Biology and Metaphor*, Cambridge University Press, Cambridge, 2002, p. 85.

25. Moore, *Nietzsche, Biology and Metaphor*, p. 108.

26. Nietzsche, quoted in R. Safranski, *Nietzsche: A Philosophical Biography*, W. W. Norton, New York, 2002, p. 200.

27. J. Richardson, *Nietzsche's New Darwinism*, Oxford University Press, Oxford, 2004; Moore, *Nietzsche, Biology and Metaphor*.

28. Nietzsche, quoted in Richardson, *Nietzsche's New Darwinism*, p. 36.

29. Descartes developed a philosophy in which the conscious, self-reflecting, thinking ego was totally separated from the body. The former was not extended in space, but the latter was. His famous epigram cogito ergo sum (I think therefore I am) has been echoed ever since by those who fail to understand or consider the embodied nature not only of our experiences but also of our cognition.

30. Richardson, *Nietzsche's New Darwinism*, p. 229.

31. For the full story see H. Ellenberger, *The Discovery of the Unconscious*, Basic Books, New York, 1981.

32. Richardson, *Nietzsche's New Darwinism*, p. 208.

33. Nietzsche, *Gay Science*, section 354, pp. 298–9.

34. *Sparagmus*: the tearing to pieces of a live animal or human.

35. C. Paglia, *Sexual Personae: Art and Decadence from Nefertiti to Emily Dickinson*, Yale University Press, New Haven, 1990, p. 96.

Chapter 2

1. Aeneas, in Virgil's *Aeneid*, l. 462.
2. Masson and McCarthy, *When Elephants Weep*; Bekoff, *Emotional Lives of Animals*. Both books are packed with observations of many animals other than humans that have been shown to behave in ways that suggest they experience an equivalent to human emotions.
3. T. Lutz, *Crying: The Natural and Cultural History of Tears*, W. W. Norton, New York, 2001.
4. *Austiefer Nothschrei ich zu dir* (from Luther, 'Eight Songs', Wittenburg, 1824).
5. 'Who passing through the valley of Baca make it a well; the rain also filleth the pools' (Psalm 84:6; King James Bible).
6. W. H. Frey, *Crying: The Mystery of Tears*, Winston Press, Minneapolis, 1985.
7. J. A. Kotter, *The Language of Tears*, Jossey-Bass, San Francisco, 1996.
8. L. Bylsma et al., 'When is Crying Cathartic? An International Study', *Journal of Social and Clinical Psychology*, 27 (2008), 1165–87.
9. Odysseus is shipwrecked on the island of the Phaeacians in the Ionian Sea, on his way home after the Trojan war. At a feast of the king, he hears Demodocus sing about Troy and Achilles.
10. Kotter, *Language of Tears*.
11. Kotter, *Language of Tears*, p. 113.
12. Anna O (Bertha Pappenheim) was a patient of Josef Breuer who suffered multiple symptoms which became referred to as hysteria. She was hypnotized by Breuer, a tale which in itself has led to much speculation about the effects the treatment had on both Anna O and Breuer. Her supposed cure was referred to as a cathartic cure, following a revived interest in the Aristotelian concept of catharsis around that time. There are many sources, but Ellenberger's *Discovery of the Unconscious* is a valuable text on the literature of exploring the unconscious and the development of psychoanalysis.
13. Kotter, *Language of Tears*, p. 94.
14. D. L. Kraemer and J. L. Hastrup, 'Crying in Natural Settings: Global Estimates, Self-Monitored Frequencies, Depression and Sex Differences in an Undergraduate Population', *Behaviour Research and Therapy*, 24 (1986), 371–3.
15. J. J. Gross, 'The Psychophysiology of Crying', *Psychophysiology*, 31 (1994), 460–8.

16. R. R. Cornelius, 'Crying and Catharsis', in A. J. J. M. Vingerhoets and R. R. Cornelius (eds.), *Adult Crying: A Biopsychosocial Approach*, Brunner-Routledge, New York, 2001, p. 204.

17. Shakespeare, *The Taming of the Shrew*, I. i. 124–6.

18. P. Ostwald, 'The Sounds of Infancy', *Developmental Medicine and Child Neurology*, 14 (1972), 350–61.

19. J. Bowlby, *Attachment and Loss*, vol. 2: *Separation, Anxiety and Anger*, Penguin, London, 1975.

20. J. Lehtonen, 'From an Undifferentiated Cry towards a Modulated Signal', *Behavioural and Brain Sciences*, 27 (2004), 467.

21. S. Laureys and S. Goldman, 'Imaging Neural Activity in Crying Infants and their Caring Parents', *Behaviour and Brain Sciences*, 27 (2004), 471–2.

22. P. MacLean, *The Triune Brain in Evolution*, Plenum Press, New York, 1990, p. 247.

23. Darwin, *Expression of the Emotions*, pp. 1351–3.

24. Darwin, *Expression of the Emotions*, pp. 1353–4.

25. Darwin, *Expression of the Emotions*, p. 1474.

26. Darwin, *Expression of the Emotions*, p. 1361.

27. A. Montague, 'Natural Selection and the Origin and Evolution of Weeping in Man', *JAMA*, 174 (1961), 130–5. Masson and McCarthy, discussing the continuing question of elephant tears in *When Elephants Weep*, note that elephant eyes water heavily, and observations of elephants crying mainly come from those who have seen this in elephants lying down. This would alter the drainage of the tears, which run when standing through nasolachrymal ducts and also down the inside of their trunks.

28. The reader is referred to D. Shapiro, *Neurotic Styles*, Basic Books, New York, 1965, for a detailed account of personality styles encountered in everyday life as well as in psychiatric practice.

29. D. G. Williams, 'Weeping by Adults: Personality Correlates', *Journal of Psychology*, 110 (1982), 217–26.

30. A. J. J. M. Vingerhoets et al., 'Personality and Crying', in A. J. J. M. Vingerhoets and R. R. Cornelius (eds.), *Adult Crying: A Biopsychosocial Approach*, Brunner-Routledge, New York, 2001, pp. 115–34; S. E. Choti et al., 'Gender and Personality Variables in Film-Induced Sadness and Crying', *Journal of Social and Clinical Psychology*, 5 (1987), 535–44.

31. Vingerhoets et al., 'Personality and Crying'.

32. Specifically, empathy increased parasympathetic tone, the other emotions increasing sympathetic activity—see Chapter 3.

33. G. Grass, *The Tin Drum*, trans. B. Mitchell. Houghton Mifflin Harcourt, New York, 2009, pp. 505–7.

34. M. Proust, *The Guermantes Way*, trans. M. Treharne, Penguin, London, 2002, p. 526.

35. W. James, *The Principles of Psychology*, Henry Holt, New York, 1890, vol. 2, p. 457.

36. P. B. Shelley, 'Music' (1824), in *The Complete Poetic Works of Percy Bysshe Shelley*. Oxford University Press, London, 1914.

37. L. Meyer, *Emotion and Meaning in Music*, University of Chicago Press, Chicago, 1956.

38. 'The practiced listener has learned to direct his attention in particular ways... hence he not only tends to improve articulation in general, but tends to favour certain types of (musical) organizations over others' (Meyer, *Emotion and Meaning*, p. 187).

39. Meyer, *Emotion and Meaning*, p. 62.

40. Meyer, *Emotion and Meaning*, p. 256.

41. Meyer, *Emotion and Meaning*, p. 258.

42. F. H. Lund, 'Why Do We Weep?', *Journal of Social Psychology*, 1 (1930), 136–51.

43. Vingerhoets et al., 'Personality and Crying', p. 79.

44. Tonality refers to a system of musical notation based on pitch variability that gives a sense that one pitch has central gravity—the tonic, the home pitch and chord of tonal music.

45. An appoggiatura is 'a kind of delay introduced to a relatively stable note by suspending it on a less stable one' (P. Ball, *The Music Instinct*, Bodley Head, London, 2010, p. 308). The cycle of fifths derives from the Pythagorean observations that the most harmonious sounds for the human ear come from combinations of pitch with the simplest ratios (e.g., 2:1, or 3:2—the perfect fifth), and nearly all known musical cultures have divided the octave into sets of pitches, duplicated in higher or lower octaves. In the Western scales, the cycle of fifths portrays relationships between the twelve tones of the chromatic scale. All the chromatic notes (the white and black notes on a piano) can be played via the repeating cycle of fifths (Ball, *Music Instinct*, p. 308).

46. J. Sloboda, 'Music Structure and Emotional Response: Some Empirical Findings', *Psychology of Music*, 19 (1991), 110–20.

47. Quoted by Lutz, *Crying*, p. 38.

48. J. Elkins, *Pictures and Tears*, Routledge, London, 2001.

49. Choti et al., 'Gender and Personality Variables'. A remake of the 1931 film, *The Champ* starred Jon Voight and Faye Dunaway.

50. Elkins, *Pictures and Tears*, p. 210.

Chapter 3

1. Frey, *Crying*.

2. V. M. W. Bodelier et al., 'Species Differences in Tears: Comparative Investigation in the Chimpanzee (*Pan Troglodytes*)', *Primates* 34 (1993), 77–84.

3. S. Gelstein et al., 'Human Tears Contain a Chemosignal', *Science*, 331 (2011), 226–30.

4. The nerves which leave the CNS through the skull. There are 12 in all and the ones of main interest to crying are the fifth maxillary, the seventh facial, and the tenth vagus (*vagus*, 'the wandering nerve').

5. For a fuller account of these historical developments see M. R. Trimble and M. George, *Biological Psychiatry*, 3rd edn, Wiley-Blackwell, Chichester, 2010.

6. For those who would prefer more, a fuller text is provided in my earlier book *The Soul in the Brain*. An even more complete journey may be found in *Biological Psychiatry*, jointly written by myself and Mark George.

7. For a full description of these methods and their limitations see Trimble and George, *Biological Psychiatry*.

8. S. A. K. Wilson, 'Some Problems in Neurology: Pathological Laughing and Crying', *Journal of Neurology and Psychopathology*, 16 (1924), 299–333.

9. A. T. Shaibani et al., 'Pathological Human Crying', in Vingerhoets and Cornelius (eds.), *Adult Crying*, pp. 265–76.

10. Amyotrophic lateral sclerosis (ALS), sometimes referred to as Lou Gehrig's disease, a form of motor neuron disease.

11. S. Arroyo et al., 'Mirth, Laughter and Gelastic Seizures', *Brain*, 116 (1993), 757–80.

12. D. Luciano et al., 'Crying Seizures', *Neurology*, 43 (1993), 2113–17. Dacrystic epilepsy is also called quitarian epilepsy, quitarian deriving from the Latin to shout loud (Quitarian was a Roman citizen).

13. William James, brother of the novelist Henry James (1843–1916). His *Principles of Psychology* was and is still considered one of the outstanding books on human psychology.

14. James, *Principles of Psychology*, vol. 2, p. 449.

15. So many guesstimates have been given, from a number greater than all the particles of sand on the beaches of the world, to one greater than all the stars in the universe—both of which are incalculable!

16. The neocortex is distinguished from older cortex by having six layers; the older cortex is variously called paleocortex or archicortex.

17. This term is used with some reluctance, but serves to hint that the circuits of the brain which are involved in the neuroanatomy of emotion are more than covered by the term 'the limbic system'. Unfortunately the latter has become fixed in the lay mind and in the descriptions of many textbooks as the part of the brain which controls emotion, without regard for the widespread representation of limbic influence and the fact that our emotions have a far greater control over our intellectual activities than many writers care to admit. This is taken up further in Chapter 6.

18. This difference in the way that smell information gets to the CNS is not properly appreciated or discussed, in part because it is assumed, quite wrongly, that smell plays a small role in human social interactions compared with the other senses.

19. Ventral means lying at the base of the nuclei, as opposed to dorsal, lying on top.

20. Korbinian Brodmann (1868–1918) divided the cerebral cortex into 52 cytoarchitectonic areas based on histological finings. Others have divided the brain up into less or more areas, but Brodmann's notation is the most widely used. In brain-imaging studies, findings are often given in relation to Brodmann areas. In this book, Brodmann areas, if they are referred to, are given in the notes. Readers who wish to see more detailed results of the data are referred to the original papers cited.

21. There were of course many others, as with any scientific journey. The author apologizes to anyone who feels that their name should be included with these three, but his own ideas have been influenced most by these neuroscientists and by his personal recollections of either working with them or discussing ideas with them during his career.

22. The discovery of these neurotransmitters in the brain, their compact organization in discrete nuclei, their chemical constitution, their receptors, and their functions has been another wonderful scientific enterprise which has revolutionized treatments in psychiatry and neurology. Selective serotonin reuptake inhibitors (SSRI) and L-dopa are just two types of drugs that have benefitted so many people with neuropsychiatric disorders from depression to Parkinson's disease. There are other transmitters, equally important to behavioural regulation, including acetylcholine (alertness and memory especially), gamma-aminobutyric acid (GABA-inhibition), and glutamate (excitation).

23. The spinal output is from sacral nerves 2 to 4. These are not related to the current discussion of emotion, but their influence over, for example, sexual function is clearly linked to emotional states.

24. About 80 percent of the fibres in the vagus nerve are afferent. This is relevant to the developing understanding of autonomic system neuroanatomy, since in the original conceptions of the autonomic systems, there was no feedback or afferent pathways (S. W. Porges, 'The Polyvagal Perspective', *Biological Psychology*, 74 (2007), 116–43).

25. Porges, 'Polyvagal Perspective'; Porges, *The Polyvagal Theory*, W. W. Norton, New York, 2011.

26. The diaphragm here being the division in the body provided anatomically by the diaphragm separating the chest from the abdomen.

27. A distinction can be drawn between the neuronal pathways controlling somatomotor structures (the muscles of the face and head) and visceromotor structures (heart, lungs etc.). Both are involved in emotional expression. Other cranial nerves involved include the ninth and the eleventh.

28. The term 'solitary nucleus' is used in this text. Its full anatomical name is the nucleus of the tractus solitarius.

29. For the interested, more details of the neuroanatomy are found in several texts, including the following: F. Leutmezer et al., 'EEG Changes at the Onset of Epileptic Seizures', *Epilepsia*, 44 (2003), 348–54; T. Deacon, *The Symbolic Species*, Allen Lane, London, 1997; L. Heimer, 'A New Anatomical Framework for Neuropsychiatric Disorders and Drug Abuse', *American Journal of Psychiatry*, 160 (2003), 1726–39; D. A. Hopkins et al., 'Vagal Efferent Projections: Viscerotopy, Neurochemistry and Effects of Vagotomy', in G. Holstege et al. (eds.), *The Emotional Motor System*,

Elsevier, Amsterdam, 1996; MacLean, *Triune Brain*; W. J. H. Nauta, 'Circuitous Connections linking Cerebral Cortex, Limbic System and Corpus Striatum', in B. K. Doane and K. E. Livingston (eds.), *The Limbic System: Functional Organization and Clinical Disorders*, Raven Press, New York, 1986, pp. 43–54; E. T. Rolls, *Emotion Explained*, Oxford University Press, Oxford, 2005; R. C. Truex and M. B. Carpenter, *Human Neuroanatomy*, 5th edn., E. and S. Livingstone, Edinburgh, 1964.

30. Deacon, *Symbolic Species*, p. 246.
31. Deacon, *Symbolic Species*, p. 431.
32. S. W. Porges, *The Polyvagal Theory*, W. W. Norton, New York, 2011.
33. Taken from Deacon, *Symbolic Species*, p. 248.
34. G. F. Striedter, *Principles of Brain Evolution*, Sinauer Associates, Sunderland, MA, 2005.
35. No other primate indicates: pointing with the index figure, a gesture with a thousand and one implications—see God's indication to Adam in Michelangelo's depiction of his creation in the Sistine Chapel. For a lovely analysis of the importance of pointing, see R. Tallis, *Michelangelo's Finger*, Atlantic Books, London, 2010.
36. A. Damasio, *The Feeling of What Happens*, Heinemann, London, 1999. This is further explored in Damasio, *Looking for Spinoza: Joy, Sorrow, and the Feeling Brain*, Heinemann, London, 2003.
37. Rolls, *Emotion Explained*.
38. S. W. Porges, 'Social Engagement and Attachment: A Phylogenetic Perspective', *Annals of the New York Academy of Sciences*, 1008 (2003), 31–47.
39. Damasio, *The Feeling of What Happens*. This is further explored in Damasio, *Looking for Spinoza: Joy, Sorrow, and the Feeling Brain*. The areas of parietal cortex he refers to are S1, S2, and medial parietal cortex.
40. For further elaboration of the animal and brain-imaging studies supporting these circuits see M. L. Phillips et al., 'The Neurobiology of Emotion Perception I: The Neural Basis of Normal Emotion Perception', *Biological Psychiatry*, 54 (2003), 504–14, and R. J. Dolan, 'Emotion, Cognition and Behaviour', *Science*, 298 (2002), 1191–4.
41. The neuroanatomy involves other subcortical structures including the peri-aqueductal grey area and the parabrachial nuclei which have been shown to be sites where direct stimulation in animals leads to emotional

expression, and coordinated responses to stress. For a fuller exposition, see Trimble and George, *Biological Psychiatry*.

42. It is positioned and often shown as a portion of the superior salivatory nucleus, situated just above the dorsal motor nucleus of the vagus in the brainstem.

43. A. Hennenlotter et al., 'The Link between Facial Feedback and Neural Activity within Central Circuitries of Emotion: New Insights from Botulinum Toxin-Induced Denervation of Frown Muscles', *Cerebral Cortex*, 19 (2009), 537–42.

44. See M. C. Corballis, *The Lopsided Ape*, Oxford University Press, Oxford, 1991.

45. The properties and functions of the non-dominant side of the brain, and their importance for regulating many aspects of human behaviour are discussed in some detail in my earlier book, *Soul in the Brain*.

46. See N. D. Cook, 'Bihemispheric Language: How the Two Hemispheres Collaborate in the Processing of Language', in T. J. Crow (ed.), *The Speciation of Modern Homo Sapiens*, Oxford University Press, Oxford, 2002, pp. 169–96; J. Cutting, *The Right Cerebral Hemisphere and Psychiatric Disorders*, Oxford University Press, Oxford, 1990; I. McGilchrist, *The Master and his Emissary: The Divided Brain and the Making of the Western World*, Yale University Press, New Haven, 2009; G. Vallortigara and L. J. Rogers, 'Survival with an Asymmetrical Brain: Advantages and Disadvantages of Cerebral Lateralization', *Behavioural and Brain Sciences*, 28 (2005), 575–89. McGilchrist's brilliant exposure of the quite extensive but diverse and well-neglected literature on the functions of the right hemisphere sounds a warning about the dangers for human civilization posed by the now overriding dominance of the dominant hemisphere in Western culture. He reveals how the functions of the two hemispheres of our bicameral brain have shifted, intermingled, conflicted, competed, and settled for us, in the 21st century, in such a skewed relationship between them that the imbalance puts our very future existence as *Homo sapiens* in considerable danger. His thesis is that at one historical time or another, the tendencies and proclivities of one hemisphere have been the more dominant, shaping the immediate culture from a historical perspective, and he acknowledges the fact that the left hemisphere has 'helped us achieve nothing less than civilization itself' (p. 93). But he thinks that we are coming close to forfeiting that very civilization.

The left hemisphere (the emissary) has taken over and dominated the right (the master), the emissary now acting like Faust, who has no insight into what is going on and cannot see the destruction in its wake.

Chapter 4

1. J. Derrida, *Of Grammatology*, Johns Hopkins University Press, Baltimore, 1997, p. 75.
2. Of an estimated 50 billion species that have perhaps graced the planet.
3. These time lines are fairly stable, but new findings are always altering the details.
4. To repeat, it is not suggested that this is the only one, but it is one human attribute which requires explanation and exploration.
5. S. Mithen, *The Singing Neanderthals*, Weidenfeld and Nicholson, London, 2005.
6. R. Burling, quoted in J. R. Hurford, *The Origins of Meaning: Language in the Light of Evolution*, Oxford University Press, Oxford, 2007, p. 183.
7. Rousseau, philosopher and composer, was best known for his writings on education, but also for developing the ideas that man is born free but is everywhere in chains, and that man is naturally good, but has been corrupted by society, civilization having bred inequality and greed.
8. Body motions express emotions, and gesture is intimately involved. Many gestures occur universally, such as a shoulder shrug and upturned palms to suggest 'no idea', and facial expressions themselves are gestural. Gestural communications are common in primates, and have been studied in apes. In humans much communication is via gesture rather than speech, unconsciously offered; to note the close links between the spoken and the gestural, simply watch someone arguing on their mobile phone when they display florid bodily movements to someone who is miles away and cannot see them! For a good account of gesture and its relation to language development see Mithen, *Singing Neanderthals*.
9. M. C. Corballis and T. Suddendorf, 'Memory, Time and Language', in C. Pasternak (ed.), *What Makes Us Human?* One world, Oxford, 2007, p. 51.
10. One feature which seems central to much activity of the human mind is generative computation, the ability to repeat and rearrange actions, words, or symbols. This allows for the creation of new forms and ideas.

The ability to adopt recursive activity by the human brain may be a feature of the recursive nature of much animal behaviour (such as walking, eating, sexual activity) later empowered by the huge increase in computational power of the six-layered laminated neocortex, and the ability to use symbols, which released the motor activities from the present to allow for a flexible future, and the use of abstract forms of cognition.

11. Primitive man living in small communities bonded socially with others. Someone close dies, but as is common for us, in dreams dead people return to us, albeit fleetingly. This must have led to an enquiry, a questioning as to where the dead were, and how it was possible to communicate with them, and to achieve such an afterlife for the dreamer. REM sleep, the periods of sleep associated with dreaming and a desynchronized EEG pattern (in contrast to slow-wave non-REM sleep), is found in all mammals.

12. J. Campbell, *The Inner Reaches of Outer Space: Metaphor as Myth and as Religion* (1986), New World Library, Novato, CA, 2002.

13. This hominid has been referred to by some as Cro-Magnon.

14. Some distinguish between *Homo sapiens* and *Homo sapiens sapiens* on the grounds that the Neanderthals should be embraced under *Homo sapiens* as well.

15. M. Ridley, *Nature via Nurture: Genes, Experience and What Makes Us Human*, Fourth Estate, London, 2003, p. 221.

16. There are many references to cremation in the Old Testament, and evidence of cremation has been found going back 40,000 years.

17. Campbell, *Inner Reaches of Outer Space*, p. 16.

18. Derrida, *Of Grammatology*, p. 31.

19. E. O. Wilson, *On Human Nature*, Harvard University Press, Cambridge, MA, 1978.

20. P. Boyer, *Religion Explained*, Heinemann, London, 2001; S. J. Blakemore, 'The Developing Social Brain: Implications for Education', *Neuron*, 65 (2010), 744–47. In the social sciences, attribution relates to determining the cause of one's own or another's behaviour.

21. R. Dawkins, *The God Delusion*, Houghton Mifflin, Boston, 2006.

22. F. de Wahl, *Primates and Philosophers: How Morality Evolved*, Princeton University Press, Princeton, 2006.

23. S. J. Carrington and A. J. Bailey, 'Are There Theory of Mind Regions in the Brain? A Review of the Neuroimaging Literature', *Human Brain Mapping*, 30 (2009), 2314.

24. R. Dunbar, 'Why are Humans Not Just Great Apes?' In Pasternak (ed.), *What Makes Us Human?*, p. 43; emphasis original.

25. Autism is a neurodevelopmental disorder, the main features of which are disturbed social interactions, repetitive stereotyped and ritualistic behaviours, and impaired speech and gesture. It is now known that a spectrum exists, which includes at the more severe end pervasive developmental disorder, and at the other Asperger's syndrome. People with these disorders do badly on Theory of Mind tests which mimic human behaviours.

26. C. D. Frith, *Making up the Mind: How the Brain Creates Our Mental World*, Blackwell, Oxford, 2007. Psychological tasks that have been developed to assess Theory of Mind involve attempts to understand stories and cartoons of social interactions, and silent animations of, for example, shapes (triangles) which seem in the animations to interact in various ways.

27. There is evidence that many animals have some idea of the future. Does not a squirrel bury its nuts? The Western scrub jay buries nuts for future consumption. Since other jays can steal their food, they seek to use spots where they cannot be observed by other jays, and some bury stones not nuts—perhaps in an attempt to mislead.

28. M. Kundera, *Immortality*, P. Harper Collins, London, 1999.

29. T. Suddendorf and M. C. Corballis, 'The Evolution of Foresight: What is Mental Time Travel, and Is It Unique to Humans?', *Behavioural and Brain Sciences*, 30 (2007), 299–351; D. L. Schacter et al., 'Remembering the Past to Imagine the Future: The Prospective Brain', *Nature Reviews: Neuroscience*, 8 (2007), 657–61.

30. The term 'medial prefrontal cortex' is somewhat imprecise, and is used differently by different authors. It includes in some definitions the anterior cingulate cortex and the frontal pole, and the areas inbetween (see D. M. Amodio and C. Frith, 'Meeting of Minds: The Medial Frontal Cortex and Social Cognition', *Nature Reviews: Neuroscience*, 7 (2006), 268–77). In terms of the Brodmann notation, it can include areas 9, 10, 11, 14, 24, 25, 32. The close links between the orbitofrontal cortex and the medial prefrontal cortex are emphasized by some using the term orbital and medial prefrontal cortex (OMPFC). See M. L. Kringelbach and E. T. Rolls, 'The Functional Neuroanatomy of the Human Orbito-Frontal

Cortex: Evidence from Neuroimaging and Neuropsychology', *Progress in Neurobiology*, 72 (2004), 341–72.

31. R. Passingham, *What is Special about the Human Brain?* Oxford University Press, Oxford, 2008; S. J. Blakemore, 'The Social Brain in Adolescence', *Nature Reviews: Neuroscience*, 9 (2008), 267–77; J. N. Giedd et al., 'Brain Development during Childhood and Adolescence: A Longitudinal Study', *Nature Neuroscience*, 2 (1999), 861–3.

32. Blakemore, 'Social Brain in Adolescence'.

33. S. G. Shamay-Tsoory, 'The Neural Basis for Empathy', *Neuroscientist*, 17 (2011), 18–24. In Brodmann this is BA 10 and BA 11.

34. Broca's area is Brodmann's area 44, but is used by some to include area 45. Paul Broca described the consequences of damage to this area on the left side of the brain, so the designation Broca's area is lateralized. However, there is the right side homologue. See Passingham, *What is Special about the Human Brain?*

35. Passingham, *What is Special about the Human Brain?*, p. 39.

36. D. R. Addis et al., 'Remembering the Past and Imagining the Future: Common and Distinct Neural Substrates during Event Construction and Elaboration', *Neuropsychologia*, 45 (2007), 1363–77.

37. D. Wildgruber et al., 'Identification of Emotional Intonation Evaluated by *f*MRI', *Neuroimage*, 24 (2005), 1233–41. Brodmann area 47.

38. McGilchrist, *Master and his Emissary*, p. 420.

39. M. Botvinick et al., 'Viewing Facial Expressions of Pain Engages Cortical Areas Involved in the Direct Experience of Pain', *Neuroimage*, 25 (2005), 312–19; T. Singer et al., 'Empathy for Pain Involves the Affective but Not Sensory Components of Pain', *Science*, 303 (2004), 1157–62.

40. Brodmann area 25.

41. Rolls, *Emotion Explained*, p. 181; J. S. Schwaber et al, 'Amygdaloid and Basal Forebrain Direct Connections with the Nucleus of the Tractus Solitarius and the Dorsal Motor Nucleus of the Vagus', *Neuroscience*, 2 (1982), 1424–38; J. L. Price, 'Connections of the Orbital Cortex', in D. H. Zald and S. L. Rauch (eds.), *The Orbitofrontal Cortex*, Oxford University Press, Oxford, 2006, pp. 39–56.

42. German -*Einfühlung* an in-feeling, something incorporated.

43. T. L. Chartrand and J. A. Bargh, 'The Chameleon Effect: The Perception–Behavior Link and Social Interaction', *Journal of Personal and Social Psychology*, 76 (1999), 893–910.

44. L. Carr et al., 'Neural Mechanisms of Empathy in Humans: A Relay from Neural Systems for Imitation to Limbic Areas', *Proceedings of the National Academy of Sciences of the United States of America*, 100 (2003), 5497–502. Premotor areas of the brain are those which will engage the direct motor area (primary motor area, Brodmann area 4, leading to the pyramidal tract output to the motor nerves) when the motor act is carried out.

45. U. Dimberg et al., 'Unconscious Facial Reactions to Emotional Facial Expressions', *Psychological Science*, 11 (2000), 86–9.

46. Carr et al., 'Neural Mechanisms of Empathy in Humans', p. 5502.

47. Rolls, *Emotion Explained*, p. 71.

48. Rolls, *Emotion Explained*, p. 62.

49. M. S. Gazzaniga, *The Mind's Past*, University of California Press, Los Angeles, 1998, p. 63.

50. B. Libet, 'Do We Have Free Will?' *Journal of Consciousness Studies*, 6 (1999), 47–57.

51. Dolan, 'Emotion, Cognition and Behaviour'.

52. M. Jeannerod, *Motor Cognition*, Oxford University Press, Oxford, 2006, p. 65.

53. D. Hume, *Treatise of Human Nature* (1739–40), Oxford University Press, Oxford, 1975, Book 2, Part 2, Section 5.

54. Silk and Stern, *Nietzsche on Tragedy*, p. 270.

55. Carrington and Bailey, 'Are There Theory of Mind Regions in the Brain?'

56. G. Buccino et al., 'Neural Circuits Involved in the Recognition of Actions Performed by Nonconspecifics: An ƒMRI Study', *Journal of Cognitive Neuroscience*, 16 (2004), 114–26.

57. B. Wicker et al., 'Both of Us Disgusted in My Insula: The Common Neural Basis of Seeing and Feeling Disgust', *Neuron*, 40 (2003), 655–64.

58. See J. V. Huxby et al., 'Human Neural Systems for Face Recognition and Social Communication', *Biological Psychiatry*, 51 (2002), 59–67.

59. Rolls, *Emotion Explained*, p. 56.

60. For further detailed information, see Rolls, *Emotion Explained*.

61. Rolls, *Emotion Explained*, p. 126.

62. A. J. Blood and R. J. Zatorre, 'Intensely Pleasurable Responses to Music Correlate with Activity in Brain Regions Implicated in Reward and Emotion', *Proceedings of the National Academy of Science*, 98 (2001), 11818–23; Rolls, *Emotion Explained*, p. 122.

63. Damasio, *Looking for Spinoza*; P. Fossati et al., 'In Search of the Emotional Self: An ƒMRI Study Using Positive and Negative Emotional Words',

American Journal of Psychiatry, 160 (2003), 1938–45; Wildgruber et al., 'Identification of Emotional Intonation Evaluated by ƒMRI'; G. R. Griffin et al., 'Theory of Mind and the Right Cerebral Hemisphere', *Laterality*, 11 (2006), 195–225; F. Benuzzi et al., 'Processing of Socially Relevant Parts of Faces', *Brain Research Bulletin*, 74 (2007), 344–56.

64. Jeannerod, *Motor Cognition*.

65. J. Searle, *Intentionality*, Cambridge University Press, Cambridge, 1983.

66. M. Merleau-Ponty, *Phenomenology of Perception*, trans. C. Smith, Routledge, London, 2002.

67. Jeannerod, *Motor Cognition*, p. 20.

68. Jeannerod, *Motor Cognition*, p. 30.

69. V. Gallese and A. Goldman, 'Mirror Neurones and the Simulation Theory of Mind Reading', *Trends in Cognitive Science*, 2 (1998), 498.

70. J. Harris, 'The Evolutionary Neurobiology, Emergence and Facilitation of Empathy', in T. F. D. Farrow and P. W. R. Woodruff (eds.), *Empathy in Mental Illness*, Cambridge University Press, Cambridge, 2006, p. 179.

71. Horace, quoted in Frey, *Crying*, p. 37.

72. Porges, *Polyvagal Theory*, pp. 202–11.

73. Corballis and Suddendorf, 'Memory, Time and Language'.

74. Ridley, *Nature via Nurture*.

75. Corballis and Suddendorf, 'Memory, Time and Language'.

76. Rolls, *Emotion Explained*, p. 57: 'Genes specify stimuli that are goals for behaviour' (p. 60), although that behavioural response is much less determined, and is subject to considerable flexibility and environmental conditioning. The genes provide goals for actions while not specifying the behaviour that leads to the accomplishment of such goals.

77. Ridley, *Nature via Nurture*. Ridley argues that the development of culture itself can select for genetic changes, allowing for the development of various dispositions (e.g., imitation, empathy, and symbolization). Nurture can express itself through nature, as the expression of genes are moderated by other genes, some for example referred to as promoters and enhancers, which switch on and off the expression of other genes. The number of genes that code for proteins in *Homo sapiens* is far fewer than was expected, much of the rest of the genome being noncoding or regulatory. This led to the silly idea that much genetic material in the human genome is 'junk DNA'. Nature is never wasteful, and

the flexibility of the genetic material is simply unknown to us at the present time.

78. Lucy (*Australopithicus afarensis*) has been upstaged by the discovery of Ardi (*Ardipithicus ramidis*), a primitive biped, thought to be 4.4 million years old.

79. Mithen, *Singing Neanderthals*, pp. 184–5; N. Wade, *Before the Dawn*, Penguin, London, 2006, p. 265.

Chapter 5

1. George Gordon, Lord Byron, *Don Juan*, Penguin, London, 1996, canto IV, st. 66.

2. The theatre of Dionysus was situated in the foothills of the Parthenon.

3. Of Aristotle's *Poetics*, only the first book survives, which deals with epic and tragic poetry. The second book on comedy is lost. Aristotle's views on Tragedy have entranced or infuriated commentators but his strictly set conditions of what constitutes Tragedy were bypassed by many later theorists and playwrights. He took an 'organic' view of the subject, constraining Tragedy with category definitions and seeking its essential properties. Aristotle went to the theatre, and, in an empirical way observed and catalogued what he saw, as he would have with a plant or an animal. For him, there were six essential ingredients to Tragedy: plot, character, diction, thought, spectacle, and melody. He was most interested in the first two, and downplayed the importance of music. His theories would seem to have caused as much academic ink being split as tears may have been shed in Tragedy. Plot—a self-contained story with a unity, and a beginning, middle, and end. J. Jones, *On Aristotle and Greek Tragedy*, Chatto and Windus, London, 1971; J. Barnes, *The Cambridge Companion to Aristotle*, Cambridge University Press, Cambridge, 1995.

4. Frazer's *Golden Bough* examined rituals and magic practices in a wide variety of cultures (J. G. Frazer, *The Golden Bough: A Study in Magic and Religion*, Macmillan, New York, 1922). He concluded that there was a natural evolutionary progression of human thought, from the magical to the religious to the scientific.

5. Frazer, *Golden Bough*.

6. The reverence for the Greeks and their culture was a German preoccupation, and had huge cultural implications. Joachim Winckelmann (1717–68) was one of a group of intellectuals who revered the beauty of

Greek art, and pronounced on a progression of styles, the classical period culminating in the golden age of Greek culture in the 5th century BC. These ideas were taken up by many others, including Johann Wolfgang von Goethe (1749–1832), the Schlegel brothers (Karl Wilhelm, 1772–1829, and August Wilhelm, 1767–1845), and Georg Hegel (1770–1831), who sought perfection in art and lamented the decline of their contemporary culture. Nietzsche too lamented the poverty of the culture of his time, but countermanded the ideas of Winckelmann by introducing the Dionysian to the Apollonian. The reaching back to an older culture to seek perfection led to Germans searching for German myths and their own past culture in order to bring about their own Renaissance. Education (*Bildung*) became a preoccupation, namely how to discipline the mind correctly, which was accelerated by the rise of Romanticism. In conjunction with German idealistic philosophies, Romanticism sought to understand the natural forces that underpin the development not only of the individual, but also of a whole culture. See P. Watson, *The German Genius*, Simon and Schuster, London, 2010, ch. 1, n. 13.

7. Thebes is the setting of *The Bacchae*, where Dionysus raises mayhem, dancing, and madness.

8. There is much in Greek literature and criticism on crying, which is beyond the scope of this book to discuss. In Greek texts women were described as crying more than men, and crying, as noted, was viewed as a womanish trait, even though some of the great heroes were given to tears. Further, crying was seen as a ruse in the Athenian courts to obtain sympathy, especially when the weeping wives of offenders were produced in court in an attempt to influence the judgment. As noted, even heroes cry. Odysseus weeps for many reasons, including the loss of his dead companions in the Trojan war. On his prolonged journey home to Ithaka, while enjoying the consolations of the lovely goddess Kalypso, he cries, for even her love and offer of immortality cannot overcome his longing to return home to his wife, Penelope. When he finally reaches home, he meets his son Telemachos, whereupon he sheds more tears. When Penelope finally recognizes him, she cries tears of joy. See B. Boyd, *On the Origin of Stories*, Belknap Press of Harvard University Press, Cambridge, MA, 2009; T. M. Falkner, 'Engendering the Tragic Theatês: Pity, Power, and Spectacle in Sophocles' *Trachiniae*', in R. H. Sternberg (ed.), *Pity and Power in Ancient Athens*,

Cambridge University Press, Cambridge, 2005, pp. 165–92; B. Hughes, *The Hemlock Cup: Socrates, Athens and the Search for the Good Life*, Jonathan Cape, London, 2010.

9. S. Brosnan and F. B. M. de Waal, 'Monkeys Reject Unequal Pay', *Nature*, 425 (2003), 297–9.

10. Bekoff, *Emotional Lives of Animals*. There is much anecdote and anthropomorphizing in these stories, which does not invalidate them, but does not allow the conclusions that the observed behaviours are altruistic in the sense that we use it in relationship to human behaviour.

11. *Illiad*, quoted in E. R. Dodds, *The Greeks and the Irrational*, University of California Press, Berkeley, 1951, p. 29.

12. J. T. Kaplan and M. Iacoboni, 'Getting a Grip on Other Minds: Mirror Neurons, Intention Understanding, and Cognitive Empathy', *Social Neuroscience*, 1 (2006), 175–83.

13. Cook, 'Bihemispheric Language', p. 169.

14. J. Jaynes, *The Origin of Consciousness in the Breakdown of the Bicameral Mind* (1976), Penguin, London, 1990.

15. Cutting, *Right Cerebral Hemisphere*; Trimble, *Soul in the Brain*; McGilchrist, *Master and his Emissary*.

16. McGilchrist, *Master and his Emissary*, p. 56.

17. McGilchrist, *Master and his Emissary*, pp. 66, 179.

18. O. Devinsky, 'Right Cerebral Hemisphere Dominance for a Sense of Corporeal and Emotional Self', *Epilepsy & Behavior*, 1 (2000), 60–73.

19. S. K. Langer, *Problems of Art*, Routledge and Kegan Paul, London, 1957, pp. 91, 133.

20. A. Tzanetou, 'A Generous City: Pity in Athenian Oratory and Tragedy', in Sternberg (ed.), *Pity and Power*, pp. 98–122.

21. Barnes, *Cambridge Companion to Aristotle*, p. 281.

22. G. Steiner, *The Death of Tragedy*, Yale University Press, New Haven, 1980, pp. 10, 164.

23. R. Scruton, *Death-Devoted Heart*, Oxford University Press, Oxford, 2004, pp. 164, 169, 180.

24. Boyd, *On the Origin of Stories*.

25. 'Man was Made to Mourn', (1785). In D. Wu (ed.), *Romanticism: An Anthology*, 3rd edn, Blackwell, Oxford, 2006, p. 266.

26. The word 'religion' comes from the Latin stem *ligare*, 'to bind'.

27. Steiner, *Death of Tragedy*, p. 8.

28. A. D. Nuttall, *Why Does Tragedy Give Pleasure?* Clarendon Press, Oxford, 1996, p. 40.

29. 'But the hand that struck my eyes was mine / mine alone—no one else—I did it all myself! / What good were eyes to me? Nothing I could see could bring me joy' (*Oedipus the King*, lines 1469–70, in Sophocles, *The Three Theban Plays*, trans. R. Fagles, Penguin, London, 1984).

Chapter 6

1. J. W. von Goethe, *Faust*, Part 1, Sc IV, 1590–1593, trans. D. Luke, Oxford World's Classics, Oxford, 1998.

2. G. Lakoff and M. Johnson, *Philosophy in the Flesh*, Basic Books, New York, 1999, p. 513.

3. Damasio, *Looking for Spinoza*, p. 128.

4. C. Finlayson, *The Humans Who Went Extinct*, Oxford University Press, Oxford, 2009, p. 217.

5. German, *fassen*, 'to take hold of, grasp'; *erfassen*, 'to comprehend, to lay hold of a matter'.

6. Damasio, *Looking for Spinoza*.

7. Hurford, *Origins of Meaning*, pp. 5–8, 173.

8. 'Nietzsche's Just So stories are terrific...aside from Nietzsche's huffing and puffing about some power subduing and becoming master, this is pure Darwin' (D. Dennett, *Darwin's Dangerous Idea: Evolution and the Meanings of Life*, Penguin, London, 1995, pp. 464, 465).

9. D. C. Dennett, *Freedom Evolves*, Penguin, London, 2003, p. 144. Ninety-five per cent of a proton's mass, and hence of you and me, comes from energy!

10. Young, *Nietzsche*, p. 221.

11. J. Young, *Nietzsche's Philosophy of Art*, Cambridge University Press, Cambridge, 1992, p. 125.

12. Young, *Nietzsche's Philosophy of Art*.

13. Silk and Stern, *Nietzsche on Tragedy*, pp. 232, 288.

14. B. L. Deputte, 'Duetting in Male and Female Songs of the White-Cheeked Gibbon (Hylobates concolor leucogenys)', in C. T. Snowdon et al. (eds), *Primate Communication*, Cambridge University Press, Cambridge, 1982, p. 80. Antiphonal, 'answering in response'.

15. D. Morris, *The Biology of Art*, Methuen, London, 1962, p. 144.

16. Acheulean: an era and industry of early stone tool-making, associated with *Homo erectus*.

17. Boyd, *On the Origin of Stories*, p. 15; Langer, *Problems of Art*.

18. The word 'projection' here is from the authors, but it is a difficult word philosophically and from the point of view of neuroscience. It begs the question as to what exactly is projected, from where, and to which destination.

19. Quoted in R. Zaretsky and J. T. Scott, *The Philosopher's Quarrel*, Yale University Press, New Haven, 2009, p. 127.

20. Nietzsche, quoted in translation in Silk and Stern, *Nietzsche on Tragedy*, p. 270.

21. S. Schachter and J. E. Singer, 'Cognitive, Social, and Physiological Determinants of Emotional State', *Psychological Review*, 69 (1962), 379–99.

22. P. F. Ferrari et al., 'The Ability to Follow Eye Gaze and its Emergence during Development in Macaque Monkeys', *Proceedings of the National Academy of Sciences*, 97 (2000), 13997–4002. A. N. Meltzoff, infant imitation and memory. *Child Development*, 59 (1988), 217–25.

23. The size of the pupil of the eye decreases with parasympathetic, and increases with sympathetic activity. It is known that atropine, by lowering parasympathetic activity, dilates the pupil, and this is thought to increase female attractiveness—*Atropa belladonna* was used by Cleopatra to achieve this effect (as if she were not beautiful enough already!). As someone with blue eyes, the author cannot help but note the fact that blue eyes are common in Europeans and their descendants. This relates to a genetic change which occurred about 6,000–10,000 years ago (G. Cochran and H. Harpending, *The 10,000 Year Explosion*, Basic Books, New York, 2009, p. 149).

24. M. Proust, *The Way by Swann's*, trans. L. Davis, Penguin, London, 2003.

25. N. A. Harrison et al., 'Processing of Observed Pupil Size Modulates Perception of Sadness and Predicts Empathy', *Emotion*, 7 (2007), 728; N. A. Harrison et al., 'Pupillary Contagion: Central Mechanisms Engaged in Sadness Processing', *Social Cognitive and Affective Neuroscience*, 1 (2006), 5–17.

26. Merleau Ponty, quoted in P. Hobson, *The Cradle of Thought*, Macmillan, London, 2002, p. 203.

27. *Tristan and Isolde*, I. ii.

28. Scruton, *Death-Devoted Heart*, p. 41.
29. Bonobo—*Pan paniscus*. Anatomically the Bonobo female vagina is situated more anteriorly compared with that of female chimpanzees.
30. Damasio, *Feeling of What Happens*, pp. 57, 80, 54.
31. C. Darwin, *On the Origin of Species by Means of Natural Selection; or The Preservation of Favoured Races in the Struggle for Life* (1859), in *From So Simple a Beginning: The Four Great Books of Charles Darwin*, ed. E. O. Wilson, W. W. Norton, New York, 2006.
32. See Darwin's section 'Struggle for Existence', in *Origin of Species*, pp. 488–500.
33. Darwin, *The Descent of Man*, in *From So Simple a Beginning: The Four Great Books of Charles Darwin*, ed. E. O. Wilson, W. W. Norton, New York, 2006, p. 1240.
34. The mechanisms whereby such changes occurred are well beyond the scope of the author and this book to discuss. However, Darwin considered that such transformations did happen, and more recent evolutionary theories employ terms like 'exaptation' to suggest the shifting functions of a trait over time. 'Adaptation' refers to natural selection shaping a character for current use, while 'exaptation' refers to a previously shaped character which is co-opted for a new use (see S. J. Gould and E. S. Vrba, 'Exaptation: A Missing Term in the Science of Form', *Paleobiology*, 8 (1982), 4–15). Gould and Vrba use the evolution of feathers in birds as one example: these were adapted for thermoregulation, then insect catching, and then were exapted for flight. 'Most of what the brain now does to enhance our survival lies in the domain of exaptation' (13). Such exaptations do not need millions of years to take place. *Homo sapiens* is but a stepping stone in primate evolution.

In *10,000 Year Explosion* Gregory Cochran and Henry Harpending consider why 21st-century mankind is still regarded as 'the epitome and termination of the evolutionary process, due to two millennia of teleological, largely religious thought, supported by a *logos*, and maintained by optimistic philosophers'. They present convincing evidence that significant evolutionary changes can be seen in as short a time space as 10,000 years, and they note how very relevant the development of agriculture was with regards to human evolution: it changed diets and social structures. Cereals and carbohydrates became a significant part of diets, and genetic mutations allowed for the digestion of lactose, the main sugar in milk. Dairy cattle provide many more calories per acre of farmed land than cattle raised for meat, and a dairy economy developed, which in turn altered social structures. 'Cultural innovation has been a driving force behind biological change in humans for a long time—certainly since the first use of tools some 2.5 million years ago. Natural selection acting on the hominid brain made those early innovations possible, and the innovations themselves led to further physical and mental changes' (p. 225).

35. Darwin, *The Expression of Emotions in Man and Animals*, in *From So Simple a Beginning: The Four Great Books of Charles Darwin*, ed. E. O. Wilson, W. W. Norton, New York, 2006, p. 1469.

36. Darwin, *Expression of Emotions*, p. 1361.

37. M. J. Magnée et al., 'Similar Facial EMG Responses to Faces, Voices, and Body Expressions', *Neuroreport*, 18 (2007), 369–72; U. Dimberg and M. Thunberg, 'Rapid Facial Reactions to Emotional Facial Expressions', *Scandinavian Journal of Psychology*, 39 (1998), 39–45.

38. A. Achaibou et al., 'Simultaneous Recording of EEG and Facial Muscle Reactions during Spontaneous Emotional Mimicry', *Neuropsychologia*, 46 (2008), 1104–13.

Chapter 7

1. I. Calvino, *If on a Winter's Night a Traveller*, Harcourt, London, 1981, p. 259.

2. Langer, *Problems of Art*, pp. 91, 133.

3. Kant seems to have considered disinterested pleasure as the aesthetic response to a work of art. Schopenhauer seems also to have considered that contemplation of works of art allowed escape from the burden of the relentless manifestations of the Will that drives the world and everything in it, the aesthetic being a disinterested experience. Needless to say Nietzsche did not agree with such notions.

4. P. J. Whalen et al., 'Human Amygdala Responses to Facial Expressions of Emotion', in P. J. Whalen and E. A. Phelps (eds.), *The Human Amygdala*, Guilford Press, New York, 2009, pp. 265–88.

5. A. R. Damasio et al., 'Subcortical and Cortical Brain Activity during the Feeling of Self-Generated Emotions', *Nature Neuroscience*, 3 (2000), 1049–56; M. S. George et al., 'Brain Activity during Transient Sadness and Happiness in Healthy Women', *American Journal of Psychiatry*, 152 (1995),

341–51; H. S. Mayberg et al., 'Reciprocal Limbic-Cortical Function and Negative Mood: Converging PET Findings in Depression and Normal Sadness', *American Journal of Psychiatry*, 156 (1999), 675–82.

6. S. Zeki, *Splendors and Miseries of the Brain*, Wiley-Blackwell, Oxford, 2009. One noted difference was the activation of the hypothalamus in the romantic love paradigm and not in that of maternal love. Zeki suggests this is connected to the sexual aspect of the former, some nuclei of the hypothalamus being involved in regulating sexual behaviours.

7. T. Ishizu and S. Zeki, 'Toward a Brain-Based Theory of Beauty', *PLoS One* 6(7), e21852, doi:10.1371/journal.pone.0021852. It should be noted that other investigations using different stimuli, such as looking at classical sculptures, have implicated the amygdala in aesthetic judgements of beauty. The story is still evolving, but the data, nonetheless, highlight the importance of this structure in our responses to different forms of art. See also C. Di Dio et al., 'The Golden Beauty: Brain Response to Classical and Renaissance Sculptures', *PLoS One*, 2(11) (2007): e1201, doi: 10.1371/journal.pone.0001201.

8. T. R. Henry et al., 'Brain Blood-Flow Alterations Induced by Therapeutic Vagus Nerve Stimulation in Partial Epilepsy: II. Prolonged Effects at High and Low Levels of Stimulation', *Epilepsia*, 45 (2004), 1064–70; A. Zobel et al., 'Changes in Regional Cerebral Blood Flow by Therapeutic Vagus Nerve Stimulation in Depression: An Exploratory Approach', *Psychiatry Research*, 139 (2005), 165–79.

9. Young, *Nietzsche*, p. 435. See also. F. Nietzsche, *Ecce Homo* (1888), trans. R. J. Hollingdale, Penguin, London, 1979, pp. 48–50.

10. Any combination of notes is a chord, and a three-note chord consisting of a note (root) and its third and fifth is called a triad. The major triad has a major third note, a minor triad has a minor one, which is the major third reduced by a semitone. It is recognized that, certainly since the 20th century, composers have shifted much musical ground, from Arnold Schoenberg's invention of the non-diatonic twelve-tone composition—atonal music, to the minimalist works of Steve Reich and Philip Glass. What is of interest is how traditional Western music has become so popular in all cultures where it has been introduced, such as India or Japan, and how musicians from such countries are so well represented in Western concert halls and opera houses. There are many good discussions of the development of music for those who are interested.

See, for example, A. Storr, *Music and the Mind*, HarperCollins, London, 1997: A. Ross, *The Rest is Noise*, Fourth Estate, London, 2007; D. Ross, *Aristotle* (1923), Routledge, London, 1996; Ball, *Music Instinct*. For those who want to explore the world of music with audio excerpts, the series of Robert Greenberg from The Teaching Company is highly recommended. WWW.TEACH12.COM.

11. R. Scruton, *Understanding Music*, Continuum, London, 2009, p. 13. Scruton observes that 'the pentatonic scale never inflicts a semitone or a tritone on the ear…even in musical traditions that avoid triadic chords the octave, forth and fifth are recognised as defining the space in which the notes of music move. There is no musical culture that I know of that does not recognise the octave as equivalent to its fundamental, and most traditions acknowledge the fifth as a metastable position on the scale, and the drone on the fifth as a stabilising harmonic accompaniment' (p. 13). The earliest musical instrument discovered, a flute, seems to have been constructed to play the pentatonic scale (see *Cave of Forgotten Dreams*, the documentary film by Werner Herzog about the art at the Chauvet caves). Flutes found in the Hohle Fels cavern in south-west Germany include one with five finger holes, and are thought to be some 40,000 years old.

12. Charles Rosen equated the sonata form (with exposition, development, and recapitulation) with dramatic action or a short story, with a beginning, middle, and end (in Storr, *Music and the Mind*, p. 81).

13. Scruton, *Understanding Music*, p. 15.

14. In the study by Blood and Zatore, 'Intensely Pleasurable Responses to Music', all the music used was Western classical, and there were no words in the pieces. Other studies have shown that when the responses of unpleasant and pleasant music are compared, the unpleasant music activates the amygdala, in contrast to pleasant music (see S. Koelsch et al., 'Investigating Emotion with Music: An *f*MRI Study', *Human Brain Mapping*, 27 (2006), 239–50). Some limbic structures, such as the parahippocampal gyrus, seem to be activated by unpleasant and dissonant music, suggesting that certain musical categories evoke different cerebral responses, which may correspond to many people's response of dislike on hearing dissonant compositions.

15. S. Koelsch, 'Investigating Emotion with Music: Neuroscientific Approaches', *Annals of the New York Academy of Science*, 1060 (2005), 412–18; Koelsch, 'Investigating Emotion with Music'.

16. Blood and Zatorre,'Intensely Pleasurable Responses to Music', p. 11823.

17. A. J. Blood et al., 'Emotional Responses to Pleasant and Unpleasant Music Correlate with Activity in Paralimbic Brain Regions', *Nature Neuroscience*, 2 (1999), 382–7; P. Janata, 'The Neural Architecture of Music-Evoked Autobiographical Memories', *Cerebral Cortex*, 19 (2009), 2579–94.

18. J. T. Hamilton, *Music, Madness and the Unworking of Language*, Columbia University Press, New York, 2008, p. 112.

19. Langer, *Problems of Art*; D. Cooke, *The Language of Music*, Oxford University Press, Oxford, 2001, p. 272.

20. The psychiatrist Anthony Reading comments: 'Hope...mobilizes us to act, to analyze and understand our problems, and to try to solve them' (*Hope and Despair*, Johns Hopkins University Press, Baltimore, 2004, p. 172). Hope indicates that we anticipate being happy in the future (p. 80). Reading takes a neurological perspective by noting how hope interlinks with the development of brain structures for our complex memory and social behaviours, especially the prefrontal cortex. He concludes by reminding us that Pandora married Prometheus, and gave him a box, in which was contained all human misery. He ill-advisedly opened it, and out flew all the evils of the world, with one exception—hope.

21. S. Baron Cohen, *Zero Degrees of Empathy: a New Theory of Human Cruelty*. Allen Lane, London, 2011.

Appendix 1

1. The terms in bold type are also referenced in Appendix 2.

BIBLIOGRAPHY

Achaibou, A., G. Pourtois, S. Schwartz, and P. Vuilleumier. 'Simultaneous Recording of EEG and Facial Muscle Reactions during Spontaneous Emotional Mimicry'. *Neuropsychologia*, 46 (2008), 1104–13.

Addis, D. R., A. T. Wong, and D. L. Schacter. 'Remembering the Past and Imagining the Future: Common and Distinct Neural Substrates during Event Construction and Elaboration'. *Neuropsychologia*, 45 (2007), 1363–77.

Adelmann, P. K. and R. B. Zajonc. 'Facial Efference and the Experience of Emotion'. *Annual Review of Psychology*, 40 (1989), 249–80.

Amodio, D. M. and C. Frith. 'Meeting of Minds: The Medial Frontal Cortex and Social Cognition'. *Nature Reviews: Neuroscience*, 7 (2006), 268–77.

Arroyo, S., R. P. Lesser, B. Gordon, et al. 'Mirth, Laughter and Gelastic Seizures'. *Brain*, 116 (1993), 757–80.

Ball, P. *The Music Instinct*. Bodley Head, London, 2010.

Barnes, J. *The Cambridge Companion to Aristotle*. Cambridge University Press, Cambridge, 1995.

Baron Cohen, S. *Zero Degrees of Empathy: A New Theory of Human Cruelty*. Allen Lane, London, 2011.

Bekoff, M. *The Emotional Lives of Animals*. New World Library, Novato, CA, 2007.

Benuzzi, F., M. Pugnaghi, S. Meletti, et al. 'Processing of Socially Relevant Parts of Faces'. *Brain Research Bulletin*, 74 (2007), 344–56.

Bindra, D. 'Weeping: A Problem of Many Facets'. *Bulletin of the British Psychological Association*, 25 (1972), 281–4.

Blakemore, S. J. 'The Social Brain in Adolescence'. *Nature Reviews: Neuroscience*, 9 (2008), 267–77.

Blakemore, S. J. 'The Developing Social Brain: Implications for Education'. *Neuron*, 65 (2010), 744–47.

Blood, A. J. and R. J. Zatorre. 'Intensely Pleasurable Responses to Music Correlate with Activity in Brain Regions Implicated in Reward and Emotion'. *Proceedings of the National Academy of Science*, 98 (2001), 11818–23.

Blood, A. J., R. J. Zatorre, P. Bermudez, and A. C. Evans. 'Emotional Responses to Pleasant and Unpleasant Music Correlate with Activity in Paralimbic Brain Regions'. *Nature Neuroscience*, 2 (1999), 382–7.

Bodelier, V. M. W., N. J. van Haeringen, and P. S. Y. Klaver. 'Species Differences in Tears: Comparative Investigation in the Chimpanzee (*Pan Troglodytes*)'. *Primates* 34 (1993), 77–84.

Botvinick, M., A. P. Jha, L. M. Bylsma, et al. 'Viewing Facial Expressions of Pain Engages Cortical Areas Involved in the Direct Experience of Pain'. *Neuroimage*, 25 (2005), 312–19.

Bowlby, J. *Attachment and Loss*, vol. 2: *Separation, Anxiety and Anger*. Penguin, London, 1975.

Boyd, B. *On the Origin of Stories*. Belknap Press of Harvard University Press, Cambridge, MA, 2009.

Boyer, P. *Religion Explained*. Heinemann, London, 2001.

Brosnan, S. and F. B. M. de Waal. 'Monkeys Reject Unequal Pay'. *Nature*, 425 (2003), 297–9.

Buccino, G., F. Lui, N. Canessa, et al. 'Neural Circuits Involved in the Recognition of Actions Performed by Nonconspecifics: An FMRI Study'. *Journal of Cognitive Neuroscience*, 16 (2004), 114–26.

Bylsma, L., A. J. J. M. Vingerhoets, and J. Rottenberg. 'When is Crying Cathartic? An International Study'. *Journal of Social and Clinical Psychology*, 27 (2008), 1165–87.

Byron, George Gordon, Lord. *Don Juan*. Penguin, London, 1996.

Calvino, I. *If on a Winter's Night a Traveller*. Harcourt, London, 1981.

Campbell, J. *The Inner Reaches of Outer Space: Metaphor as Myth and as Religion* (1986). New World Library, Novato, CA, 2002.

Carr, L., M. Iacoboni, M. C. Dubeau, et al. 'Neural Mechanisms of Empathy in Humans: A Relay from Neural Systems for Imitation to Limbic Areas'. *Proceedings of the National Academy of Sciences of the United States of America*, 100 (2003), 5497–502.

Carrington, S. J. and A. J. Bailey. 'Are There Theory of Mind Regions in the Brain? A Review of the Neuroimaging Literature'. *Human Brain Mapping*, 30 (2009), 2313–35.

Chartrand, T. L. and J. A. Bargh. 'The Chameleon Effect: The Perception–Behavior Link and Social Interaction'. *Journal of Personal and Social Psychology*, 76 (1999), 893–910.

Choti, S. E., A. R. Marston, S. G. Holston, and J. T. Hart. 'Gender and Personality Variables in Film-Induced Sadness and Crying'. *Journal of Social and Clinical Psychology*, 5 (1987), 535–44.

Cochran, G. and H. Harpending. *The 10,000 Year Explosion*. Basic Books, New York, 2009.

Cook, N. D. 'Bihemispheric Language: How the Two Hemispheres Collaborate in the Processing of Language'. In T. J. Crow (ed.), *The Speciation of Modern Homo Sapiens*. Oxford University Press, Oxford, 2002, 169–96.

Cooke, D. *The Language of Music*. Oxford University Press, Oxford, 2001.

Corballis, M. C. *The Lopsided Ape*. Oxford University Press, Oxford, 1991.

Corballis, M. C. and T. Suddendorf. 'Memory, Time and Language'. In C. Pasternak (ed.), *What Makes Us Human?* Oneworld, Oxford, 2007, 17–36.

Cornelius, R. R. 'Crying and Catharsis'. In A. J. J. M. Vingerhoets and R. R. Cornelius (eds.), *Adult Crying: A Biopsychosocial Approach*. Brunner-Routledge, New York, 2001, 199–211.

Cutting, J. *The Right Cerebral Hemisphere and Psychiatric Disorders*. Oxford University Press, Oxford, 1990.

Damasio, A. *The Feeling of What Happens*. Heinemann, London, 1999.

Damasio, A. *Looking for Spinoza: Joy, Sorrow, and the Feeling Brain*. Heinemann, London, 2003.

Damasio, A. R., T. J. Grabowski, A., et al. 'Subcortical and Cortical Brain Activity during the Feeling of Self-Generated Emotions'. *Nature Neuroscience*, 3 (2000), 1049–56.

Darwin C. *The Descent of Man (1871). In From So Simple a Beginning: The Four Great Books of Charles Darwin*, ed. E. O. Wilson. W. W. Norton, New York, 2006.

Darwin, C. *The Expression of the Emotions in Man and Animals (1872). In From So Simple a Beginning: The Four Great Books of Charles Darwin*, ed. E. O. Wilson. W. W. Norton, New York, 2006.

Darwin, C. *On the Origin of Species by Means of Natural Selection; or The Preservation of Favoured Races in the Struggle for Life (1859). In From So Simple a Beginning: The Four Great Books of Charles Darwin*, ed. E. O. Wilson. W. W. Norton, New York, 2006.

Dawkins, R. *The God Delusion*. Houghton Mifflin, Boston, 2006.

Deacon, T. *The Symbolic Species*. Allen Lane, London, 1997.

Dennett, D. *Darwin's Dangerous Idea: Evolution and the Meanings of Life*. Penguin, London, 1995.

Dennett, D. C. *Freedom Evolves*. Penguin, London, 2003.

Deputte, B. L. 'Duetting in Male and Female Songs of the White-Cheeked Gibbon (Hylobates concolor leucogenys)', in C. T. Snowdon et al. (eds), *Primate Communication*, Cambridge University Press, Cambridge, 1982, 67–93.

Derrida, J. *Of Grammatology*. Johns Hopkins University Press, Baltimore, 1997.

Derrida, J. *Writing and Difference*. Routledge, London, 1978.

Devinsky, O 'Right Cerebral Hemisphere Dominance for a Sense of Corporeal and Emotional Self'. *Epilepsy & Behavior*, 1 (2000), 60–73.

De Wahl, F. *Primates and Philosophers: How Morality Evolved*. Princeton University Press, Princeton, 2006.

Di Dio, C., E. Macaluso, and G. Rizzolatti. 'The Golden Beauty: Brain Response to Classical and Renaissance Sculptures'. *PLoS One* 2(11) (2007): e1201, doi: 10.1371/journal.pone.0001201.

Dimberg, U. and M. Thunberg. 'Rapid Facial Reactions to Emotional Facial Expressions'. *Scandinavian Journal of Psychology*, 39 (1998), 39–45.

Dimberg, U., M. Thunberg, and K. Elmehed. 'Unconscious Facial Reactions to Emotional Facial Expressions'. *Psychological Science*, 11 (2000), 86–9.

Dodds, E. R. *The Greeks and the Irrational*. University of California Press, Berkeley, 1951.

Dolan, R. J. 'Emotion, Cognition and Behaviour'. *Science*, 298 (2002), 1191–4.

Dunbar, R. 'Why are Humans Not Just Great Apes?' In C. Pasternak (ed.), *What Makes Us Human?* Oneworld, Oxford, 2007, 37–48.

Elkins, J. *Pictures and Tears*. Routledge, London, 2001.

Ellenberger, H. *The Discovery of the Unconscious*. Basic Books, New York, 1970.

Euripides. *Bacchae*, trans. J. Morwood. Oxford University Press, Oxford, 2000.

Falkner, T. M. 'Engendering the Tragic Theatês: Pity, Power, and Spectacle in Sophocles' Trachiniae', in R. H. Sternberg (ed.), *Pity and Power in Ancient Athens*. Cambridge University Press, Cambridge, 2005, 165–92.

Ferrari, P. F., E. Kohler, L. Fogassi, and V. Gallese. 'The Ability to Follow Eye Gaze and its Emergence during Development in Macaque Monkeys'. *Proceedings of the National Academy of Sciences*, 97 (2000), 13997–4002.

Finlayson, C. *The Humans Who Went Extinct*. Oxford University Press, Oxford, 2009.

Fossati, P., S. J. Hevenor, S. J. Graham, et al. 'In Search of the Emotional Self: An *f* MRI Study Using Positive and Negative Emotional Words'. *American Journal of Psychiatry*, 160 (2003), 1938–45.

Fossey, D. *Gorillas in the Mist*. Houghton Mifflin, Boston, 1983.

Frazer, J. G. *The Golden Bough: A Study in Magic and Religion*, Macmillan, New York, 1922.

Freud, S. and J. Breuer. *Studies in Hysteria*, vol. 3. Penguin, London, 1974.

Frey, W. H. *Crying: The Mystery of Tears*. Winston Press, Minneapolis, 1985.

Frith, C. D. *Making up the Mind: How the Brain Creates Our Mental World*. Blackwell, Oxford, 2007.

Gallese, V. and A. Goldman. 'Mirror Neurones and the Simulation Theory of Mind Reading'. *Trends in Cognitive Science*, 2 (1998), 493–501.

Gazzaniga, M. S. *The Mind's Past*. University of California Press, Los Angeles, 1998.

Gelstein, S., Y. Yeshurun, L. Rozenkrantz, et al. 'Human Tears Contain a Chemosignal'. *Science*, 331 (2011), 226–30.

Gentilucci, M. and M. C. Corballis. 'The Hominid that Talked'. In C. Pasternak (ed.), *What Makes Us Human?* Oneworld, Oxford, 2007, 49–70.

George, M. S., T. A. Ketter, P. I. Parekh, et al. 'Brain Activity during Transient Sadness and Happiness in Healthy Women'. *American Journal of Psychiatry*, 152 (1995), 341–51.

Giedd, J. N., J. Blumenthal, N. O. Jeffries, et al. 'Brain Development during Childhood and Adolescence: A Longitudinal Study'. *Nature Neuroscience*, 2 (1999), 861–3.

Gould, S. J. and E. S. Vrba. 'Exaptation: A Missing Term in the Science of Form'. *Paleobiology*, 8 (1982), 4–15.

Grass, G. *The Tin Drum*, trans. B. Mitchell. Houghton Mifflin Harcourt, New York, 2009.

Griffin, G. R., O. Friedman, E. Winner, et al. 'Theory of Mind and the Right Cerebral Hemisphere'. *Laterality*, 11 (2006), 195–225.

Gross, J. J. 'The Psychophysiology of Crying'. *Psychophysiology*, 31 (1994), 460–8.

Hamilton, J. T. *Music, Madness and the Unworking of Language*. Columbia University Press, New York, 2008.

Harris, J. 'The Evolutionary Neurobiology, Emergence and Facilitation of Empathy'. In T. F. D. Farrow and P. W. R. Woodruff (eds.). *Empathy in Mental Illness.* Cambridge University Press, Cambridge, 2006, 168–86.

Harrison, N. A., C. E. Wilson, and H. D. Critchley. 'Processing of Observed Pupil Size Modulates Perception of Sadness and Predicts Empathy'. *Emotion,* 7 (2007), 724–9.

Harrison, N. A., T. Singer, P. Rotshtein, et al. 'Pupillary Contagion: Central Mechanisms Engaged in Sadness Processing'. *Social Cognitive and Affective Neuroscience,* 1 (2006), 5–17.

Heimer, L. 'A New Anatomical Framework for Neuropsychiatric Disorders and Drug Abuse'. *American Journal of Psychiatry,* 160 (2003), 1726–39.

Heimer, L., G. W. Van Hoesen, M. Trimble, and D. S. Zahm. *Anatomy of Neuropsychiatry.* Elsevier/Academic Press, Burlington, MA, 2008.

Hennenlotter, A., C. Dresel, F. Castrop, et al. 'The Link between Facial Feedback and Neural Activity within Central Circuitries of Emotion: New Insights from Botulinum Toxin-Induced Denervation of Frown Muscles'. *Cerebral Cortex,* 19 (2009), 537–42.

Henry, T. R., R. A. Bakay, P. B. Pennell, et al. 'Brain Blood-Flow Alterations Induced by Therapeutic Vagus Nerve Stimulation in Partial Epilepsy: II. Prolonged Effects at High and Low Levels of Stimulation'. *Epilepsia,* 45 (2004), 1064–70.

Hobson, P. *The Cradle of Thought.* Macmillan, London, 2002.

Hopkins, D. A., D. Biger, J. de Vente J, and H. W. M. Steinbusch. 'Vagal Efferent Projections: Viscerotopy, Neurochemistry and Effects of Vagotomy'. In G. Holstege, R. Bandler, and C. B. Saper (eds.), *The Emotional Motor System.* Elsevier, Amsterdam, 1996.

Hughes, B. *The Hemlock Cup: Socrates, Athens and the Search for the Good Life.* Jonathan Cape, London, 2010.

Hume, D. *Treatise of Human Nature* (1739–40). Oxford University Press, Oxford, 1975.

Hurford, J. R. *The Origins of Meaning: Language in the Light of Evolution.* Oxford University Press, Oxford, 2007.

Huxby, J. V., E. A. Hoffman, and M. I. Gobbini. 'Human Neural Systems for Face Recognition and Social Communication'. *Biological Psychiatry,* 51 (2002), 59–67.

Iacoboni, M. 'Imitation, Empathy, and Mirror Neurons'. *Annual Review of Psychology,* 60 (2009), 653–70.

Iacoboni, M., R. P. Woods, M. Brass, et al. 'Cortical Mechanisms of Human Imitation'. *Science*, 286 (1999), 2526–8.

Ishizu, T. and S. Zeki. 'Toward a Brain-Based Theory of Beauty'. *PLoS One* 6(7), e21852, doi:10.1371/journal.pone.0021852.

James, W. *The Principles of Psychology*. Henry Holt, New York, 1890.

Janata, P. 'The Neural Architecture of Music-Evoked Autobiographical Memories'. *Cerebral Cortex*, 19 (2009), 2579–94.

Jaynes, J. *The Origin of Consciousness in the Breakdown of the Bicameral Mind* (1976). Penguin, London, 1990.

Jeannerod, M. *Motor Cognition*. Oxford University Press, Oxford, 2006.

Jones, J. *On Aristotle and Greek Tragedy*. Chatto and Windus, London, 1971.

Joyce, J. *A Portrait of the Artist as a Young Man* (1916). Signet Classics, New York, 1991.

Kaplan, J. T. and M. Iacoboni. 'Getting a Grip on Other Minds: Mirror Neurons, Intention, Understanding, and Cognitive Empathy'. *Social Neuroscience*, 1 (2006), 175–83.

Kaufmann, W. *Tragedy and Philosophy*. Princeton University Press, Princeton, 1968.

Kobayashi, C., G. H. Glover, and E. Temple. 'Children's and Adults' Neural Bases of Verbal and Non-Verbal "Theory of Mind"'. *Neuropsychologia*, 45 (2007), 1522–32.

Koelsch, S. 'Investigating Emotion with Music: Neuroscientific Approaches'. *Annals of the New York Academy of Science*, 1060 (2005), 412–18.

Koelsch, S., T. Fritz, D. Y. v. Cramon, et al. 'Investigating Emotion with Music: An fMRI Study'. *Human Brain Mapping*, 27 (2006), 239–50.

Koestler, A. *The Act of Creation*. Penguin, London, 1964.

Kotter, J. A. *The Language of Tears*. Jossey-Bass, San Francisco, 1996.

Kraemer, D. L. and J. L. Hastrup. 'Crying in Natural Settings: Global Estimates, Self-Monitored Frequencies, Depression and Sex Differences in an Undergraduate Population'. *Behaviour Research and Therapy*, 24 (1986), 371–3.

Kringelbach, M. L. and E. T. Rolls. 'The Functional Neuroanatomy of the Human Orbito-Frontal Cortex: Evidence from Neuroimaging and Neuropsychology'. *Progress in Neurobiology*, 72 (2004), 341–72.

Kundera, M. *Encounter Essays*. Faber and Faber, London, 2011.

Kundera, M. *Immortality*. HarperCollins, London, 1999.

Lakoff, G. and M. Johnson. *Philosophy in the Flesh*. Basic Books, New York, 1999.

Langer, S. K. *Problems of Art*. Routledge and Kegan Paul, London, 1957.

Laureys, S. and S. Goldman. 'Imaging Neural Activity in Crying Infants and their Caring Parents'. *Behaviour and Brain Sciences*, 27 (2004), 471–2.

Lehtonen, J. 'From an Undifferentiated Cry towards a Modulated Signal'. *Behavioural and Brain Sciences*, 27 (2004), 467.

Leutmezer, F., C. Schernthaner, S. Lurger, et al. 'EEG Changes at the Onset of Epileptic Seizures'. *Epilepsia*, 44 (2003), 348–54.

Libet, B. 'Do We Have Free Will?' *Journal of Consciousness Studies*, 6 (1999), 47–57.

Liébert, G. *Nietzsche and Music*. University of Chicago Press, Chicago, 2004.

Luciano, D., O. Devinsky, and K. Perrine. 'Crying Seizures'. *Neurology*, 43 (1993), 2113–17.

Lund, F. H. 'Why Do We Weep?' *Journal of Social Psychology*, 1 (1930), 136–51.

Lutz, T. *Crying: The Natural and Cultural History of Tears*. W. W. Norton, New York, 2001.

MacLean, P. *The Triune Brain in Evolution*. Plenum Press, New York, 1990.

Magee, B. *The Story of Philosophy*. Dorling Kindersley, New York, 2001.

Magnée, M. J., J. J. Stekelenburg, C. Kemner, and B. de Gelder. 'Similar Facial EMG Responses to Faces, Voices, and Body Expressions'. *Neuroreport*, 18 (2007), 369–72.

Masson, J. and S. McCarthy. *When Elephants Weep: The Emotional Life of Animals*. Vintage, New York, 1996.

Mayberg, H. S., M. Liotti, S. K. Brannan, et al. 'Reciprocal Limbic-Cortical Function and Negative Mood: Converging PET Findings in Depression and Normal Sadness'. *American Journal of Psychiatry*, 156 (1999), 675–82.

McGilchrist, I. *The Master and his Emissary: The Divided Brain and the Making of the Western World*. Yale University Press, New Haven, 2009.

Meltzoff, A. N. 'Infant Imitation and Memory: Nine-Month-Olds in Immediate and Deferred Tests'. *Child Development*, 59 (1988), 217–25.

Merleau-Ponty, M. *Phenomenology of Perception*, Trans. C. Smith. Routledge, London, 2002.

Meyer, L. *Emotion and Meaning in Music*. University of Chicago Press, Chicago, 1956.

Mithen, S. *The Singing Neanderthals*. Weidenfeld and Nicholson, London, 2005.

Montague, A. 'Natural Selection and the Origin and Evolution of Weeping in Man'. *JAMA*, 174 (1960), 392–7.

Moore, G. *Nietzsche, Biology and Metaphor*. Cambridge University Press, Cambridge, 2002.

Morris, D. *The Biology of Art*. Methuen, London, 1962.

Nauta, W. J. H. 'Circuitous Connections linking Cerebral Cortex, Limbic System and Corpus Striatum'. In B. K. Doane and K. E. Livingston (eds.), *The Limbic System: Functional Organisation and Clinical Disorders*. Raven Press, New York, 1986, 43–54.

Neafsey, E. J., R. R. Terreberry, K. M. Hurley, et al. 'Anterior Cingulate Cortex in Rodents: Connections, Visceral Control Functions, and Implications for Emotion'. In B. A. Vogt and M. Gabriel (eds.), *Neurobiology of Cingulate Cortex and Limbic Thalamus*. Birkhäuser, Boston, 1993, 206–23.

Nietzsche, F. *The Birth of Tragedy: Out of the Spirit of Music*, trans. S. Whiteside. Penguin, London, 1993.

Nietzsche, F. *Ecce Homo* (1888), trans. R. J. Hollingdale. Penguin, London, 1979.

Nietzsche, F. *The Gay Science* (1881), trans. W. Kaufmann. Vintage, New York, 1974.

Nuttall, A. D. *Why Does Tragedy Give Pleasure?* Clarendon Press, Oxford, 1996.

Ostwald, P. 'The Sounds of Infancy'. *Developmental Medicine and Child Neurology*, 14 (1972), 350–61.

Paglia, C. *Sexual Personae: Art and Decadence from Nefertiti to Emily Dickinson*. Yale University Press, New Haven, 1990.

Panksepp, J. *Affective Neuroscience: The Foundations of Human and Animal Emotions*. Oxford University Press, Oxford, 1998.

Papez, J. W. 'A Proposed Mechanism of Emotion'. *Archives of Neurology and Psychiatry*, 38 (1937), 725–43.

Passingham, R. *What is Special about the Human Brain?* Oxford University Press, Oxford, 2008.

Phillips, M. L., W. C. Drevets, S. L. Rauch, and R. Lane. 'The Neurobiology of Emotion Perception I: The Neural Basis of Normal Emotion Perception'. *Biological Psychiatry*, 54 (2003), 504–14.

Pinker, S. *The Language Instinct*. Penguin, London, 1994.

Porges, S. W. 'The Polyvagal Perspective'. *Biological Psychology*, 74 (2007), 116–43.

Porges, S. W. *The Polyvagal Theory*. W. W. Norton, New York, 2011.

Porges, S. W. 'Social Engagement and Attachment: A Phylogenetic Perspective'. *Annals of the New York Academy of Sciences*, 1008 (2003), 31–47.

Price, J. L. 'Connections of the Orbital Cortex'. In D. H. Zald and S. L. Rauch (eds.), *The Orbitofrontal Cortex*. Oxford University Press, Oxford, 2006, 39–56.

Proust, M. *The Guermantes Way*, trans. M. Treharne. Penguin, London, 2002.

Proust, M. *The Way by Swann's*, trans. L. Davis. Penguin, London, 2003.

Raphael, D. D. *The Paradox of Tragedy*. Allen and Unwin, London, 1960.

Reading, A. *Hope and Despair*. Johns Hopkins Press, Baltimore, 2004.

Richardson, J. *Nietzsche's New Darwinism*. Oxford University Press, Oxford, 2004.

Rickard, N. S. 'Intense Emotional Responses to Music: A Test of the Physiological Arousal Hypothesis'. *Psychology of Music*, 32 (2004), 371–88.

Ridley, M. *Nature via Nurture: Genes, Experience and What Makes Us Human*. Fourth Estate, London, 2003.

Rizzolatti, G. *Mirrors in the Brain: How our Minds Share Actions and Emotions*. Oxford University Press, Oxford, 2006.

Rizzolatti, G., L. Fogassi, and V. Gallese. 'Mirrors in the Mind'. *Scientific American*, 295 (2006), 30–7.

Rolls, E. T. *Emotion Explained*. Oxford University Press, Oxford, 2005.

Ross, A. *The Rest is Noise*. Fourth Estate, London, 2007.

Ross, D. *Aristotle* (1923). Routledge, London, 1996.

Safranski, R. *Nietzsche: A Philosophical Biography*. W. W. Norton, New York, 2002.

Sallis, J. 'Shining Apollo'. In W. Santaniello (ed.), *Nietzsche and the Gods*. SUNY Press, New York, 2001, 57–72.

Scammell, M. *Koestler: The Indispensable Intellectual*. Faber and Faber, London, 2009.

Schachter, S. and J. E. Singer. 'Cognitive, Social, and Physiological Determinants of Emotional State'. *Psychological Review*, 69 (1962), 379–99.

Schacter, D. L., D. R. Addis, and R. L. Buckner. 'Remembering the Past to Imagine the Future: The Prospective Brain'. *Nature Reviews: Neuroscience*, 8 (2007), 657–61.

Schopenhauer, A. *The World as Will and Idea*. J. M. Dent, London, 1995.

Schwaber, J. S., B. S. Kapp, G. A. Higgins, and P. R. Rapp. 'Amygdaloid and Basal Forebrain Direct Connections with the Nucleus of the Tractus Solitarius and the Dorsal Motor Nucleus of the Vagus'. *Neuroscience*, 2 (1982), 1424–38.

Scruton, R. *Death-Devoted Heart*. Oxford University Press, Oxford, 2004.

Scruton, R. *Understanding Music*. Continuum, London, 2009.

Searle, J. *Intentionality*. Cambridge University Press, Cambridge, 1983.

Shaibani, A. T., N. Sabbagh, and B. N. Khan. 'Pathological Human Crying'. In J. J. M. Vingerhoets and R. R. Cornelius (eds.), *Adult Crying: A Biopsychosocial Approach*. Brunner-Routledge, New York, 2001, pp. 265–76.

Shamay-Tsoory, S. G. 'The Neural Basis for Empathy'. *Neuroscientist*, 17 (2011), 18–24.

Shapiro, D. *Neurotic Styles*. Basic Books, New York, 1965.

Shelley, P. B. *The Complete Poetic Works of Percy Bysshe Shelley*. Oxford University Press, London, 1914.

Silk, M. S. and J. P. Stern. *Nietzsche on Tragedy*. Cambridge University Press, Cambridge, 1981.

Singer, T., B. Seymour, J. O'Doherty, et al. 'Empathy for Pain Involves the Affective but not Sensory Components of Pain'. *Science*, 303 (2004), 1157–62.

Sloboda, J. 'Music Structure and Emotional Response: Some Empirical Findings'. *Psychology of Music*, 19 (1991), 110–20.

Sophocles. *The Three Theban Plays*, trans. R. Fagles. Penguin, London, 1984.

Steiner, G. *The Death of Tragedy*. Yale University Press, New Haven, 1980.

Sternberg, R. H. (ed.). *Pity and Power in Ancient Athens*. Cambridge University Press, Cambridge, 2005.

Storr, A. *Music and the Mind*. HarperCollins, London, 1997.

Striedter, G. F. *Principles of Brain Evolution*. Sinauer Associates, Sunderland, MA, 2005.

Suddendorf, T. and M. C. Corballis. 'The Evolution of Foresight: What is Mental Time Travel, and Is It Unique to Humans?' *Behavioural and Brain Sciences*, 30 (2007), 299–351.

Tallis, R. *Michelangelo's Finger*. Atlantic Books, London, 2010.

Trimble, M. R. *The Soul in the Brain: The Cerebral Basis of Language, Art, and Belief*. Johns Hopkins University Press, Baltimore, 2007.

Trimble, M. R. and M. George. *Biological Psychiatry*, 3rd edn. Wiley-Blackwell, Chichester, 2010.

Truex, R. C. and M. B. Carpenter. *Human Neuroanatomy*, 5th edn. E. and S. Livingstone, Edinburgh, 1964.

Turner, S. 'The Universe in the Beginning'. *Scientific American* (Sept. 2009), 21–9.

Tzanetou, A. 'A Generous City: Pity in Athenian Oratory and Tragedy', in R. H. Sternberg (ed.), *Pity and Power in Ancient Athens*. Cambridge University Press, Cambridge, 2005.

Vallortigara, G. and L. J. Rogers. 'Survival with an Asymmetrical Brain: Advantages and Disadvantages of Cerebral Lateralization'. *Behavioural and Brain Sciences*, 28 (2005), 575–89.

Vingerhoets, A. J. J. M., M. A. L. Van Tilburg, A. J. W. Boelhouwer, and G. L. Van Heck. 'Personality and Crying'. In A. J. J. M. Vingerhoets and R. R. Cornelius (eds.), *Adult Crying: A Biopsychosocial Approach*. Brunner-Routledge, New York, 2001, pp. 115–34.

Vingerhoets, A. J. J. M. and R. R. Cornelius (eds.). *Adult Crying: A Biopsychosocial Approach*. Brunner-Routledge, New York, 2001.

Wade, N. *Before the Dawn*. Penguin, London, 2006.

Watson, P. *The German Genius*. Simon and Schuster, London, 2010.

Whalen, P. J., F. C. Davis, J. A. Oler, et al. 'Human Amygdala Responses to Facial Expressions of Emotion'. In P. J. Whalen and E. A. Phelps (eds.), *The Human Amygdala*. Guilford Press, New York, 2009, pp. 265–88.

Wicker, B., C. Keysers, and J. Plailly, et al. 'Both of Us Disgusted in My Insula: The Common Neural Basis of Seeing and Feeling Disgust'. *Neuron*, 40 (2003), 655–64.

Wilde, O. *Salome*. Nick Hern, London, 2010.

Wildgruber, D., A. Riecker, I. Hertrich, et al. 'Identification of Emotional Intonation Evaluated by fMRI'. *Neuroimage*, 24 (2005), 1233–41.

Williams, D. G. 'Weeping by Adults: Personality Correlates'. *Journal of Psychology*, 110 (1982), 217–26.

Wilson, E. O. *On Human Nature*. Harvard University Press, Cambridge, MA, 1978.

Wilson, S. A. K. 'Some Problems in Neurology: Pathological Laughing and Crying'. *Journal of Neurology and Psychopathology*, 16 (1924), 299–333.

Wong, K. 'The Morning of the Modern Mind'. *Scientific American*, 292 (2005), 86–95.

Wu, D. (ed.). *Romanticism: An Anthology*, 3rd edn. Blackwell, Oxford, 2006.

Young, J. *Friedrich Nietzsche: A Philosophical Biography*. Cambridge University Press, Cambridge, 2010.

Young, J. *Nietzsche's Philosophy of Art*. Cambridge University Press, Cambridge, 1992.

Zaretsky, R. and J. T. Scott. *The Philosopher's Quarrel*. Yale University Press, New Haven, 2009.

Zeki, S. *Splendors and Miseries of the Brain*. Wiley-Blackwell, Oxford, 2009.

Zobel, A., A. Joe, N. Freymann, et al. 'Changes in Regional Cerebral Blood Flow by Therapeutic Vagus Nerve Stimulation in Depression: An Exploratory Approach'. *Psychiatry Research*, 139 (2005), 165–79.

INDEX

Page number in **bold** indicate a glossary entry